Galvanic Batteries: Their Theory, Construction and Use

Primary, Single and Double Filled Cells, Secondary and Gas Batteries

by S.R. Bottone

with an introduction by Roger Chambers

Self Reliance Books

Get more historic titles on animal and stock breeding, gardening and old
fashioned skills by visiting us at:

http://selfreliancebooks.blogspot.com/

Introduction

I am pleased to present yet another title on Historical Science.

The work is in the Public Domain and is re-printed here in accordance with Federal Laws.

As with all reprinted books of this age that are intended to perfectly reproduce the original edition, considerable pains and effort had to be undertaken to correct fading and sometimes outright damage to existing proofs of this title. At times, this task is quite monumental, requiring an almost total "rebuilding" of some pages from digital proofs of multiple copies. Despite this, imperfections still sometimes exist in the final proof and may detract from the visual appearance of the text.

I hope you enjoy reading this book as much as I enjoyed making it available to readers again.

Roger Chambers

PREFACE

ALTHOUGH the dynamo has ousted the battery from its position as chief generator of current electricity, there still remains a large field in which the use of the dynamo is inadmissible. Even when the dynamo is applicable, cases may occur wherein the employment of a battery is preferable, owing either to local circumstances, or to considerations of the time for which the current is required. Letters innumerable have been received from correspondents respecting the suitability of the many batteries which have been brought before the public, since the original discoveries of Galvani and Volta, for special purposes. The following chapters embody the substance of the replies to these queries; and the result is a work, in which a description of every known battery of any practical use is given; along with data as to E.M.F., internal resistance, and adaptability to particular requirements. In order to render the work useful from a scientific point of view, as well as under the practical aspect, the theory of the battery has been carefully gone into; and formulæ showing the reactions that go on in the different types of cells are inserted. By this means, the knowledge acquired will be found at once comprehensive and of practical utility.

S. Bottone.

Wallington, 1902.

CONTENTS

ILLUSTRATIONS

ILLUSTRATIONS

INDEX TO SECTIONS

INDEX TO SECTIONS

INDEX TO SECTIONS

INDEX TO SECTIONS

GALVANIC BATTERIES

CHAPTER I

THE DISCOVERIES OF SULZER, GALVANI, VOLTA, ETC.

§ 1. THE knowledge of several isolated facts due to the excitation of electricity by mechanical means, such as friction, percussion, cleavage, etc., had gradually been accumulating from the time when Thales, 300 B.C., had observed the production of attractive effects by rubbing amber, down to the period when Dr. Gilbert published in 1600 his very original and valuable work on the magnet and kindred subjects, wherein an attempt at connecting these facts was made. But it was not until 1767 that we find any notice of phenomena referable to the manifestation of electricity due to mere contact of different metals or to chemical action. In that year there occurs in a metaphysical work, entitled *The General Theory of Pleasure*, by a German writer of the name of Sulzer, the record of an observation, that the application of two dissimilar metals, one placed above and the other below the tongue, gave rise to the perception of a peculiar taste when the two free ends of the metals were brought into contact.

B

He ascribed this sensation to *some vibratory motion*, excited by the contact of the metals, and communicated to the nerves of the tongue. "Content with this loose and fanciful explanation" (says Dr. Roget in his treatise on Galvanism, published in 1831), "Sulzer appears to have pursued the subject no further." In the present state of our knowledge it would seem that this *fanciful explanation* is the correct one; and that all the manifestations which we are accustomed to classify under the name of "electricity," are due simply to vibratory motion, either molecular or atomic.

§ 2. It happened, in the year 1790, that the wife of Galvani, who was professor of anatomy at Bologna, was advised to take, as a nutritive article of diet, some soup made of the flesh of frogs. Several of these creatures, recently skinned for that purpose, were lying on a table in the laboratory, close to an electrical machine, with which a pupil of the professor was amusing himself in trying experiments. While the machine was in action, he chanced to touch the bare nerve of the leg of one of the frogs with the blade of a knife which he held in his hand, when suddenly the whole limb was thrown into violent convulsions. Galvani was not present when this occurred, but received the account from his lady, who had witnessed, and had been struck by the singularity of the appearance. He lost no time in repeating the experiment, in examining carefully all the circumstances connected with it, and in determining those on which its success depended. He ascertained that the convulsions took place only at the moment that the spark was drawn from

the prime conductor, and the knife was at the same time
in contact with the nerve of the frog. He next found
that other metallic bodies might be substituted for the
knife; and very justly inferred that they owed this
property of exciting muscular contractions to their being
good conductors of electricity.

§ 3. Far from being satisfied with having arrived at this
conclusion, it only served to stimulate him to the further
investigation of this curious subject; and his perseverance
was at length rewarded by the discovery, that similar
convulsions might be produced in a frog, independently
of the electrical machine, by forming a chain of conducting
substances between the outside of the muscles of the leg
and the crural nerve. Galvani had previously entertained
the idea that the contractions of the muscles of animals
were in some way dependent on electricity; and as the
new experiments appeared strongly to favour this hypothesis, he with great ingenuity applied it to explain them.
He compared the muscles of a living animal to a Leyden
phial, charged by the accumulation of electricity on its
surface; while he conceived that the nerve belonging to
it performed the function of the wire communicating with
the interior of the phial, which would of course be charged
negatively. In this case whenever a communication was
made, by means of a substance of high conducting power,
between the surface of the muscle and the nerve, the
equilibrium would be instantly restored, and a sudden
contraction of the fibres would be the consequence.

§ 4. The discoveries of Galvani were no sooner made
known to the scientific world, than they excited very

general interest; and philosophers in every country of
Europe vied with each other in repeating his experiments,
in varying them in all possible ways, and in inventing all
kinds of hypotheses to account for the phenomena.   Some
regarded them as the effects of some new and unknown
agent, differing altogether from electricity; while others
adopted the views of Galvani, recognized them to be
electrical, and attributed them to a peculiar modification
of the power residing in the animal system only, and
which they accordingly distinguished by the name of
Animal Electricity.

But the discovery of new facts contributed more and
more to multiply and strengthen the analogies between
"galvanism" and electricity; till at length all doubt of
the identity of the agent concerned in all these phenomena
was removed by the discovery of the Galvanic or Voltaic
Pile.   Whatever share accident may have had in the
original discovery of Galvani, it is certain that the invention of the pile, an instrument which has most
materially contributed to the extension of our knowledge
in this branch of physical science, was purely the result
of reasoning.   Professor Volta of Pavia, a name already
familiar to electricians, was led to the discovery of its
properties by deep meditation on the development of
electricity at the surface of contact of different metals.
We may justly regard this discovery as forming an
important epoch in the history of Galvanism; and indeed
since that period, the terms Voltaism, or Voltaic Electricity,
have often, in honour of this illustrious philosopher, been
used to designate that particular form of electrical agency,

instruments for the liberation of which form the subject of the present work.

§ 5. The means adopted for obtaining manifestations of galvanic electricity was to interpose between two plates of different kinds of metal, a fluid capable of exerting some chemical action on one of the plates, and then to establish a communication between the plates at some other part, either by their direct contact with one another,

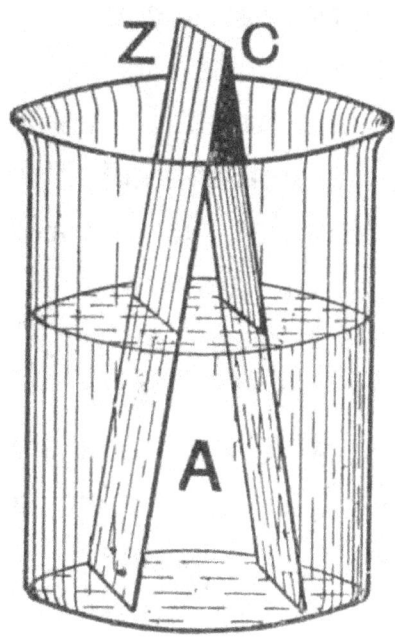

Fig. 1.

or by the intervention of some conducting substance.   Let us take, for example, a plate of zinc Z, and another of copper C, and immerse them, to a certain depth only, in diluted sulphuric acid A, contained in a glass vessel, keeping their lower edges at a little distance from one another; then, inclining them towards each other, let us bring their upper edges (which are out of the fluid) into contact, as shown in our Fig. 1.   The arrangement we have just formed constitutes what is called a galvanic circle in the simplest form, of which the three parts, or elements, are zinc, acid and copper; each of these bodies being in contact with the two others. Under these circumstances it is found that an electrical disturbance is set up; a current passing from the zinc to the acid, from the acid to the copper, from the copper back again to the zinc, and so on in a continuous circuit.

§ 6. The same effect will take place if, instead of allowing the metallic plates to come into direct contact, a communication between them be effected by wires extending from the one to the other. The circuit of electricity will thus be lengthened, but the current will move in the same direction as before, namely, positive electricity flowing from the zinc through the fluid towards the copper, and thence along the wire, from the copper, back to the zinc, as shown in Fig. 2. The completion of the circuit by means of a wire enables us to direct the electric current through such bodies as we may wish to subject to its influence, and at the same time gives us the power of interrupting or renewing at pleasure the communication between the two metallic plates by merely separating or joining together their remote extremities. When united, the copper plate is imparting electricity to the wire which touches the zinc plate, hence the former is considered as being in a positive and the latter in a negative state.

§ 7. The electrical effects of the simple apparatus just described are, in general, too feeble to be readily perceived, unless by certain delicate tests. The fact mentioned by

Sulzer, and the experiments of Galvani on the muscles of frogs, in their original form, afford however examples of the operation of simple galvanic circles. When the tongue is interposed between zinc and copper, or zinc and silver, the saliva in contact with the metals acts the part of an acid in attacking the zinc, and the current of electricity

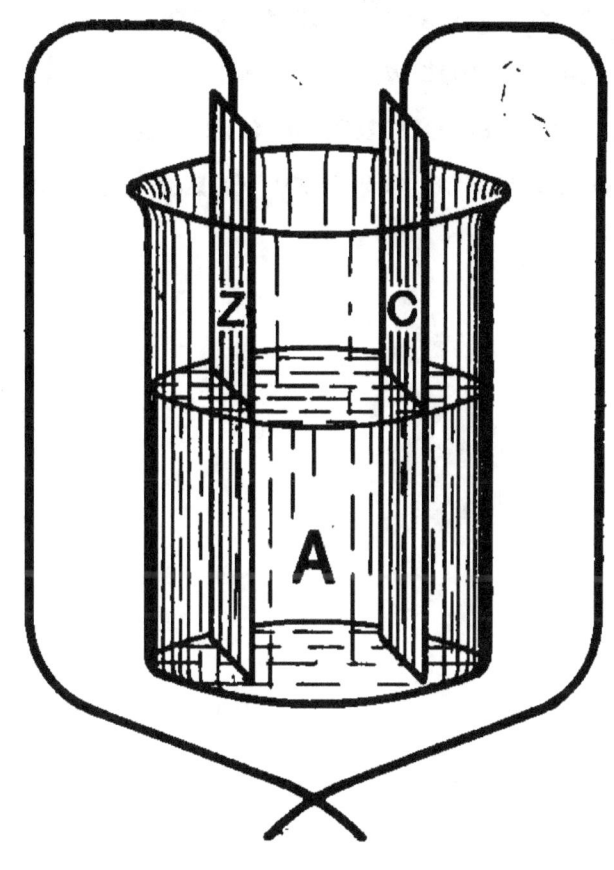

Fig. 2.

thus set up, in its passage from the zinc to the copper, through the substance of the tongue, affects the nerves of that organ, so as to give rise to the sensations of taste, which may be heightened by the decomposition of the saliva itself under the influence of the electric current. In Galvani's experiments, muscular contractions were produced by forming a connection between two different metals, one of which was applied to the nerve, and the other to the muscles of a frog's leg. It is evident that such an arrangement constitutes a galvanic circuit, deriving its activity from the chemical action of the fluids in the frog's leg, on the more oxidizable of the two metals with which they are in contact. Although the quantity of electricity set up by this slight action is necessarily very minute, yet it is sufficient when passing over the exquisitely sensitive nerves of the tongue, or the highly irritable fibres of the frog, to produce a very considerable impression thereon.

CHAPTER II

CAUSES OF THE PHENOMENA

§ 8. BEFORE proceeding to describe the many modifications and improvements which have been suggested or effected in batteries since the discoveries mentioned in the last chapter, it will be well for us to consider the theories propounded to account for the results; as by so doing we shall be enabled to obtain a better grasp of principles which are involved in the production of a good battery. It must not be supposed that any of the theories are free from objections, or that the one now generally accepted may not be capable of considerable improvement; but rather that a knowledge of these theories, based as they are on the observation and comparison of many known facts, must afford us a better insight into the probable results of any given combination than could be obtained by the consideration of isolated facts.

§ 9. What electricity is we really know not. According to the most plausible views, it would appear to be one of the forms of energy (similar to light and heat) inherent to matter; and to manifest itself when by some particular means motion is imparted to the atoms of bodies. Hence we may conveniently define electricity as being " a mode of motion in the atoms of bodies." Whether this motion be vibratory, rotary, or otherwise, we have at present no means of determining. In many of its manifestations and effects it is closely allied to, if not identical with, light and heat; and in its mode of propagation along matter and in space obeys the same general laws. It bears, in many of its aspects, a considerable resemblance to that coarser vibratory motion in the mass of bodies to which we give the name *sound.* One thing is certain: these three manifestations of energy, heat, light and electricity are mutually convertible. (Under the heading electricity we include *magnetism,* which is but another aspect of electricity.) From the above considerations, it will be immediately understood that we cannot produce or generate electricity, since every atom of every known body is endowed with a certain definite and specific amount of energy, fixed and invariable for each particular element, which energy we can neither produce nor destroy, but of which we can only cause a transference. In point of fact, as Professor Oliver Lodge well puts it, " You are to think of an electric machine as a pump, which, being attached to two bodies respectively, drives some electricity from the one into the other, conferring upon the one a positive and upon the other a precisely equal negative charge."

§ 10. We have already noticed, in § 4, that Galvani, while recognizing the results he obtained as being electrical, rather tended to the belief (at least at first) that the source of the electrical disturbance resided in the muscles and nerves of the living animal; while Volta, basing his reasoning on the results of several experiments instituted to this end, came to the conclusion that the effects were due, not to the material lying between the two dissimilar metals, but primarily to the mere contact of the two metals. It had long been suspected, rather than proved, that a feeble excitation of electricity was set up by the contact of different metals; but the fact was not clearly established until Volta, about the year 1801, devised means of proving experimentally this result. The apparatus he employed in his investigations on this subject consisted of two discs,

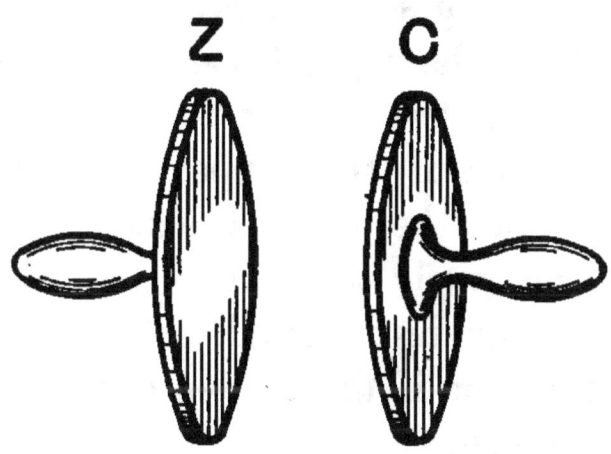

Fig. 3.

one of zinc and one of copper (Fig. 3), rather more than two inches in diameter, ground perfectly plane, and furnished with insulating handles at their centres, these being perpendicular to their surfaces. By means of these handles the plates could be brought into contact without being touched with the hand, or any conductor. With this precaution the discs were caused to approach till they touched one another; they were then separated, care being taken to keep them parallel during separation. The electricity they possessed after this separation was then examined by means of a condenser; and in order that the effects might be rendered more evident, the electrical disturbance set up by a number of successive contacts and separations was accumulated on the same condenser, care being taken to restore the discs to the neutral condition after each contact. It was constantly found that the copper disc charged the condenser with negative electricity, while the zinc disc as constantly gave it a positive charge) Thus it was established as a general fact, that these two metals, insulated, and in their natural state, are brought, by simple contact, into opposite electrical conditions, the zinc acquiring a positive charge, and the copper becoming to an equal extent negatively charged. The same result can be obtained even if the copper and zinc discs be soldered together, and the one or the other surface of the discs be made to touch the condensing plate without separation. Here the result could not be due to either friction or pressure; for if the condenser plate (which was of copper) were touched with the zinc disc, the copper disc to which it is attached being held in the hand, no trace of electricity is observed. On these results Volta founded his "contact theory," and laid down the general law, that "when two heterogeneous substances are placed in contact, one of them always assumes the positive and the other the negative electrical condition." Hence arose a memorable controversy between Galvani and Volta. Galvani, as we have already seen, held to the opinion that the source of the electrical disturbance lay, not in the metals employed, but in the substance placed between the metals; and subsequent discoveries, while weakening his argument as to the presence of "animal electricity," served to strengthen his contention that the metals employed acted principally as conductors of the current set up.   Dr. Baconio of Milan composed a galvanic pile entirely of vegetable substances, namely, discs of red beetroot two inches in diameter, and discs of walnut-tree of the same diameter, these latter having been freed from their resinous matter by digestion in a solution of cream of tartar in vinegar.   With a pile so constructed, using a leaf of scurvy grass as a conductor, he is said to have excited convulsive movements in the legs of a frog.[1]   Some of the experiments of Aldini, which were performed in the presence of a committee appointed by the French Institute, and then repeated with success in London, at the Anatomical Theatre, Great Windmill Street, tended still farther to strengthen Galvani's views on this subject.   Aldini succeeded in producing galvanic convulsions in the muscles of animals recently killed, without the intervention of any metallic substance, sometimes by bringing into contact the nerve of one animal with the muscle of another, and in other cases by employing the nerves and muscles of the same animal.   In some of his experiments the most powerful contractions were excited by bringing the parts of a warm-blooded animal into contact with those of a cold-blooded animal.   On introducing, for instance, into one of the ears of an ox recently killed a finger of one hand (previously moistened with salt water), and holding in the other hand a prepared frog, when the spine of the frog was made to touch the tongue of the ox, convulsions took

---

[1] *Nicholson's Journal*, xviii. 159.

place in the limb of the frog.　In like manner, when he held a prepared frog in his hand, moistened with a solution of salt, and applied the crural nerves of the batrachian to his own tongue, convulsions were set up.[1]　Similar experiments with more refined apparatus have been since instituted by Nobili, Matteucci, Donders, and Du Bois Reymond, and have greatly extended our knowledge of this subject.　We now know that although electricity is the *cause* of the effects noticed, both in the case of batteries consisting in alternations of different metals, and in the case of the juxtaposition of vegetable and recently-killed animal muscles and nerves, yet the source of the electricity is quite different.　It was long before the contact theory of Volta was finally abandoned, so plausible did his reasoning seem.　But it soon met with considerable opposition. Fabroni, a compatriot of Volta, having noticed that the discs of zinc which were employed in the "pile" became oxidized on the surface, considered that this oxidization was the chief factor in the production of the electrical disturbance.　The experiments of Messrs. Nicholson and Carlisle, Henry, Haldane, Sir Humphry Davy, conduced more and more to strengthening this belief, which led up to the so-called "chemical theory" of the action in the battery.　De La Rive showed, that although in Volta's fundamental experiment signs of electricity were certainly obtained, yet if the zinc be held by a wooden clamp, or if the experiment be performed in an atmosphere of any gas incapable of acting chemically on the zinc, such as hydrogen or nitrogen, no disturbance of electrical equilibrium

[1] *Nicholson's Journal*, iii. 298.

is set up. From this he concluded that the results obtained by Volta were due to the chemical action of the perspiration, and to the action of the oxygen in the atmosphere on the zinc. That chemical actions *do* set up

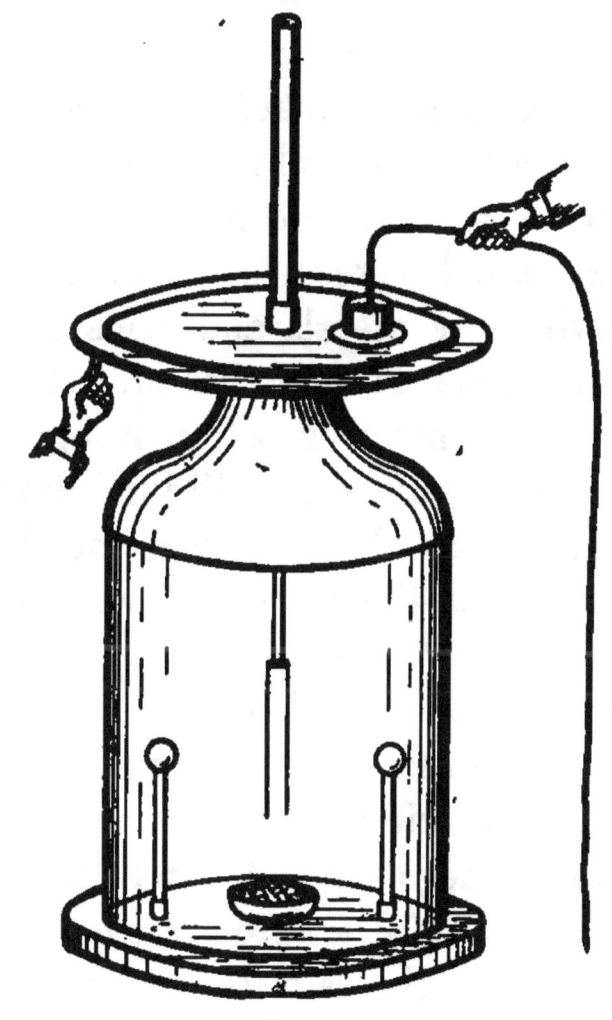

Fig. 4.

this electrical disturbance may be easily demonstrated by the aid of a condensing electroscope (Fig. 4). A disc of moistened paper is placed on the upper plate of the condenser. Resting on this is placed a zinc cup containing a little dilute sulphuric acid (which acts chemically upon

zinc, dissolving it, with the formation of zinc sulphate).
A platinum wire, which must be insulated from the zinc
cup, is immersed in the dilute acid, which has no action
on the platinum. This latter is placed in electrical
communication with the ground. Things being in this
position, the lower plate of the condenser is now put in
connection with the ground by touching with the moistened
finger. Contact being now broken, and the upper plate of
the condenser removed, the gold leaves will be found to
diverge and show positive electrification, which proves
that the upper plate has received a negative charge.

§ 11. By varying the metals and acids employed, and
by other modifications, it may be proved that various
chemical actions are instrumental in bringing about this
disturbance in the electrical equilibrium, the most active
being those which take place between metals and liquids
capable of acting upon them. Hence we may summarize
the chemical theory in these words :—

" When chemical action takes place between a liquid
and a metal, a disturbance in the electric equilibrium takes
place, the liquid acquiring a positive condition, and the
metal an equivalent negative state, and the electro-motive
force thus displayed is directly proportional to the intrinsic
energy of the metal (in calorics), minus the opposing
energy of the liquid employed."

It is true that Lord Kelvin (then Sir Wm. Thompson)
showed [1] that an electric disturbance can be set up by two
dissimilar metals in contact, whereas no such result is
obtained when these two are connected by a drop of

---

[1] *Papers on Electrostatics*, p. 317. Macmillan and Co., about 1873.

water; but this can hardly be considered an absolute proof, since the resistance of the water is so high as to minimize the resulting current, if any. The Thompson experiment was conducted as follows:—A very light metal bar is freely suspended, so as to be able to oscillate in the horizontal plane. A ring made in two halves, one being of zinc and the other of copper, and retained in contact either by clamping or by soldering, was placed so as to encircle one extremity of the light bar. On electrifying this bar negatively, it turned from the copper to the zinc half; but if it were positively electrified it turned from the zinc to the copper. This would tend to prove that the mere contact of the two metals caused them to assume opposite electrical conditions, the zinc becoming positive and the copper negative. If, instead of being in metallic contact, the two halves of the ring were united by a drop of water, the light bar would not alter its position relatively to the ring (provided of course that it hung symmetrically as regards the two halves of the ring) whether it were electrified positively or negatively. Whether this can be looked upon as a convincing proof that the movement of electricity in the galvanic circuit is entirely due to metallic contact is more than doubtful. We may therefore regard it as highly probable, that although metallic contacts *may* be capable of giving rise to electric disturbance, nay, may even bring about chemical action, yet in the cell or battery the greater part of the electro-motive force is due to the chemical action of the fluid on the metal acted on. The reader will find much instructive matter bearing on this question, along with a description of additional experiments favouring the chemical theory, in a paper by J. A. Fleming, published in the Proceedings of the Physical Society (Taylor and Francis). The pages of the *Electrical Review* between the years 1895 and 1900 teem with letters and descriptions of experiments, tending to prove, or at least to render almost certain, the truth of the chemical theory of the cause of setting up of the galvanic or voltaic effect in the battery.

CHAPTER III

EXPLANATION OF TECHNICAL TERMS, ELECTRO-MOTIVE FORCE, RESISTANCE, CURRENT, ETC.

§ 12. In our last chapter we employed the term "electro-motive force." As in speaking of batteries, frequent use will be made of this expression, it will be desirable to define it somewhat, and also to point out on what it depends in the case particularly under consideration. Unfortunately, the term is an erroneous one, since, strictly speaking, it is not a force at all; at least not in the Newtonian sense, in which *force* is only that which acts on *matter*. The definitions usually given for electro-motive force are :—

1st. "The cause which sets up currents of electricity."

2nd. "The force starting electricity in motion."

3rd. "The force which moves or tends to move electricity."

4th. Prof. Sylvanus P. Thompson defines the term as follows :—"The term electro-motive force is employed to denote *that which moves or tends to move electricity from one place to another.*"

We shall employ the term (which will hereafter be usually abbreviated as E.M.F.) to express whatever cause tends to upset the electrical equilibrium in a body, or a circuit of bodies, this disturbance naturally tending to produce a flow or current of electricity from the body, which for the time being has in any manner received a higher charge of electricity than the adjacent body or bodies, towards that body or those bodies which have a lower charge. Viewed under this aspect, it will be seen that it may be roughly compared to any cause which produces a difference of level in liquids. Suppose, for example, we have

Fig. 5.

two equal cisterns filled to the same level with water, and we connect them below by a short tube. It will be evident that under these conditions there will be no flow, and no tendency of flow, of the water in *A* (Fig. 5) into *B*, or *vice versâ*, since the pressure is the same and equal. But if by any means we disturb this equilibrium, or balance of pressure, or "potential," say by pumping water out of *A* into *B*, or from *B* into *A*, there will be an immediate tendency on the part of the water in the vessel in which the water stands at the higher level, to flow through the connecting pipe into the vessel in which the water stands at the lower level; and this tendency will result, provided the connection be kept open, in a continuous flow of water from the former vessel to the latter, until the level has again been reached, when, the pressures or "potential" being equal, all flow will cease. It will readily be understood, that if by any means we can keep up a difference in level between the water in the two cisterns (while they are connected below, either by pumping into the one, or abstracting water from the other), we shall obtain in the connecting tube, a flow or *current*, which will continue as long as the "water-motive force" is applied.

§ 13. In applying this to electro-motive force as we study it in the galvanic cell, we must bear in mind the statement made in Chapter II, that the atoms of the elements are endowed with a certain definite amount[1] of *energy* or *force*, linked to, or charged upon the atoms of these elements, or upon the molecules, if they are in a state of combination. John T. Sprague, referring to this specific energy in his work *Electricity*, says, " This is what used to be called *latent heat*, now termed, particularly by mathematicians, *potential energy;* it is inseparable from the atoms or molecules of matter without change of nature or of physical state, the mode of charging being the imparting of internal motion. Every elementary atom has its special amount of force, and the importance of this

---

[1] Invariable for each particular element, but different for different elements.

view will be seen when we find that the degree of this force is really the measure and the cause of the chemical force or affinities of this atom. Each substance also requires a definite force to pass from one physical state to a higher, as from a solid to a liquid or a gaseous; at each such change a definite amount of force disappears, becomes charged on the molecules, *i. e.* is converted into latent heat or potential energy. So also *every chemical action* which occurs under the influence of affinity, that is every act of *combination*, is attended with a loss of force, *i. e.* potential energy is set free, and becomes active and sensible in *some* form, either as heat, or *electricity*, or motion. On the other hand, every act of *decomposition* (the reversal of affinity) requires a supply of force exactly equal in quantity to that set free by the act of combination, and this force is again charged upon the atoms, and disappears—without it the change cannot occur." On careful consideration of the above, it will be understood that during the act of chemical combination, let us say, for example, the combustion of one grain of carbon in oxygen (with the production of carbon dioxide), we find that a certain definite amount of energy is set free in the form of *heat*, sufficient to raise the temperature of 1 lb. of water through nearly 12·47 degrees Fahr. Converting this into the well-known mechanical units " foot lbs.," we find that this is equivalent to 9624[1] foot lbs. If zinc be used instead of carbon, the energy set free in the form of heat will be found

---

[1] These numbers must not be regarded as absolutely correct, they are near approximations; but different experimenters have obtained slightly different results.

to be equal to 8607·5 foot lbs. per grain of zinc combining with oxygen. Hence we may fairly consider these figures as representing the *specific energy* of carbon and of zinc respectively. In like manner, it has been found by actual experiment that one grain of copper burning in oxygen to form cupric oxide, develops 3802 foot lbs.; hydrogen 6726; iron 6565; lead 5494; potassium 15135; silver 1214; sodium 14593; tin 6654, etc. It will be evident from this, that if our assumption be correct (based as it is on the correlation of forces), that every element has a specific amount of energy inherent to its atom or molecule, we can foretell or calculate the E.M.F. which a given combination of elements may elicit. It will hereafter be seen, when we come to treat of the actual E.M.F. displayed by well-known batteries, the chemical reactions going on in which have been carefully studied, that our assumption is verified in practice. We can therefore state with certainty, that although the mere contact may be, and is actually attended by the development of a small amount of electromotive force, yet the principal factor concerned in this result in the Voltaic battery is the chemical action taking place between the liquid and one of the metals or "elements" constituting the battery, and that the measure of E.M.F. set up depends entirely upon the "energy of combination" or "affinity" between the metal acted upon and the liquid employed to excite this action. The necessary components of a battery therefore are: 1st, a metal capable of entering into chemical combination with a liquid, thereby liberating a given amount of energy; 2nd, a liquid (acid or otherwise) capable of acting on this first metal; and 3rd, a metal or other conductor, which is either not acted on at all by the said acid or liquid, or is acted on by it to a much lesser extent. This act of chemical combination (like that of combustion in oxygen) liberates the latent energy, thereby producing a disturbance in the electric equilibrium, this being equivalent to a difference of potential, or an E.M.F., causing a current of electricity to flow from the first plate to the second, through the fluid, thence back again to the first plate (provided connection be established between the two plates), as long as chemical action can go on between the first plate and the acid or other existing liquid.

§ 14. We are now in a position to understand that it is perfectly possible to *measure* this electro-motive force (*i. e.* to give it a value as compared to some other forms of energy), and to study its effects. It is extremely unfortunate that, owing to the multiplicity of units of weights, measures, etc., which we are still using in this country, and the want of relation between these and the units of force, that it is almost impossible, unless we have recourse to the French metric system, to obtain anything like a satisfactory idea of the relation which exists between force, velocity, mass, weight, and measure. And even with the metric system, several points that are of the highest scientific interest, such as, for instance, the real relation which exists between the chemical equivalent of an element, its valency, and its intrinsic energy, are left without any satisfactory means of being linked to the other well-known units. We shall, however, here endeavour to show how these different measures are related, and hence demonstrate that it is possible to deduce the actual value in foot-pounds, or in "calories" of the "volt," which is the name given to the unit of electro-motive force. To understand this, it will be necessary to remember that all the physical quantities, such as *force, velocity,* etc., can be expressed by their relation to *length, mass,* and *time.* For instance, we can define that amount of force which we call 1 *horse power,* as being the force necessary to raise 33,000 pounds (mass) through 1 foot (length) in 1 minute (time). It is not necessary here to go very deeply into all the units of measurement; but in order to be able to grasp the relationship that exists between the length, time, and mass (or weight) units, and the ones used in electrical science (and which are derived from these), it will be well for us to give a brief *résumé* of the system of units, which has received the almost universal sanction of the scientific world. This is called the "Centimetre-Gramme-Second" (generally referred to as the "C.G.S." system).

§ 15. In this, the C.G.S. system, the *Centimetre* is taken as the unit of length, the *Gramme,* which is a cubic centimetre of water at 4° Centigrade, is at once the unit of mass and of weight (since its weight is also called 1 gramme—equal to 15·432 grains), and the *Second,* as the unit of time. Now from these three fundamental units are deduced many others, for instance, *area* can be measured in square centimetres, *volume* in cubic centimetres, the *unit of velocity* may be defined as a speed of 1 centimetre in 1 second Again, *force* may be measured by the velocity that it will impart to mass in a given time. The unit of force is called a *Dyne,* and it is that amount of force which, acting for 1 second on a mass of 1 gramme, will impart to it a unit of velocity; that is to say, will cause it to move 1 centimetre in 1 second. The unit of *work* is that amount of work done in pushing or pulling a body against a force of 1 dyne, through a distance of 1 centimetre. It is called the *Erg*. The name given to the unit of *Energy* is also *Erg*, since it is evident we can only measure the energy contained in a body by the *work* it will do.

That particular form of force or energy which we call *heat*, is usually measured by the amount required to raise the temperature of 1 gramme mass (1 cubic centimetre) of water from 0 to 1 degree Centigrade. It is called 1 *Caloric*. The force or energy expended in order to do this is found to be 42,000,000 ergs. From the above-mentioned units, it is easy to derive a set of units for electrical measurements. Thus we may define a current to be of *unit strength*, when 1 centimetre of its length, bent into an arc of 1 centimetre radius, shall exert a force of 1 dyne on a unit magnetic pole placed in the centre. In like manner the unit quantity of electricity may be defined as being that quantity which is conveyed by a unit current in one second. As the *unit of potential* or electro-motive force it is usual to take the expenditure of one erg of work to bring a unit of positive electricity from one point to the other against the electric force.[1] Finally, we can measure the resistance which a body presents to the passage of electricity, and define the *unit of resistance* as being that amount of resistance which

---

[1] Opposing force.

requires one unit of potential applied to it to cause a current of unit strength (that is to say, one unit of quantity per second) to flow through it.

§ 16. For practical work, these electrical C.G.S. units, or "absolute" units as they are also called, are found in some cases to be inconveniently small, and in others as much too large. For example, there is hardly a galvanic cell known, the E.M.F. of which does not amount to several millions of the above described absolute or C.G.S. units of electro-motive force; take, for instance, a zinc-copper couple immersed in dilute sulphuric acid, and we find it gives above 100,000,000 of these C.G.S. units, so that to express the E.M.F. of the Daniell cell, we should be compelled to write it as 107,000,000. Again, in the measurement of current strength, it is found in practice that few cells give a current of unit strength; hence it has been found convenient to use multiples or sub-multiples of these "absolute" or C.G.S. units, in actual work. For this reason the *practical* unit of electro-motive force, which is called the VOLT, is 100,000,000 C.G.S. units; the measure of current is the AMPÈRE, which is $\frac{1}{10}$ of the C.G.S. unit, as defined above; and lastly, the measure of resistance now adopted is the OHM, which is 1,000,000,000 absolute or C.G.S. units.

§ 17. Another means of connecting these three latter units with those of *weight,* is to be found in the fact that when a chemical compound is decomposed into its component elements by means of a definite quantity of electricity in a given time, the amounts of each element thus separated are definite and proportional to the "equivalent"[1] of each particular element; for example, if we place in circuit with one another a number of troughs containing respectively sulphate of copper, cyanide of silver, sulphate of iron, sulphate of zinc, etc., and cause to pass through all the troughs simultaneously such a quantity of electricity that shall in one second liberate 31·5 grammes of copper, we shall find that in the same time, and by the same current, there will have been set free 108 grammes of silver, 28 grammes of iron, 32·5 grammes of zinc, etc. In fact, one celebrated scientist[2] has given the name "chemic" to that quantity of electricity which is capable of liberating 1 grain of hydrogen (from its combinations) in 10 hours. In the C.G.S. system the "chemic" would be that quantity of current electricity that would liberate 1 gramme of hydrogen in 1 second (this would be equal to 9659·97 absolute units, or to 96599·7 ampères per second very nearly). The practical use of this unit, the "chemic," is that it enables us at a glance to see, first, the weight in grammes of any given element that will be liberated per second by that unit; secondly, it gives us an immediate and definite means of ascertaining the amount of force or "energy" which can be liberated by any element during the act of combination; this energy displaying itself either as heat, as electricity, etc., according to circumstances. As in electro-chemical actions, such as in the battery and in the decomposition cell, the *atomic weights* and the *equivalents* of the elements

---

[1] Not atomic weight, though in many instances the two are the same. See Table of Atomic Weights and Electro-chemical Equivalents.

[2] J. T. Sprague.

do not always coincide (owing to the different *valencies* or combining powers of the different bodies), we give here a table of those elements with which we shall have more particularly to deal, as they enter as integral parts, in some one or other of the different forms of Voltaic cells or batteries, hereinafter to be described.

§ 18. Table of Elements, with their atomic weights, valencies, electric equivalents, etc.

Name.	Symbol.	Atomic weight.	Valency.	Electro-chemical Equivalent.	Intrinsic Energy [1] in Ergs, per gramme.	Energy in [2] Ft. lbs., per grm.
Hydrogen	H	1	1	1	14270513 millions	6800
Zinc	Zn	65·2	2	32·6	18039167 ,,	8600
Iron (as fer-rosum)	Fe	56	2	28	15761627 ,,	7510
Nickel	Ni	59	2	29·5		
Lead	Pb	207	2	103·5	11529683 ,,	5494
Tin	Sn	118	2	59·	13963985 ,,	6654
Copper (as cupricum)	Cu	63·5	2	31·75	10168207 ,,	4802
Silver	Ag	108·	1	108·	2547885 ,,	1214
Platinum	Pt	197	4	49·25		
Carbon	C	12	4	...	{ 6182923 as CO	2946
					20198389 as CO_2	9624
Mercury	Hg	200	2	100		
Manganese	Mn	55	2 and 6	27·5		

[1] Intrinsic energy expressed in ergs, for one gramme equivalent burnt in oxygen.
[2] Intrinsic energy expressed in foot pounds, per grain equivalent burnt in oxygen. Where no number appears, trustworthy data are not to hand.

§ 19. We now see whence this energy which tends to set up a current, and which we call E.M.F., is derived, and what its amount may be, according to the nature of the chemical action which goes on between the exciting fluid and the active element in the cell. But we do not gather from the above statements any idea of the actual quantity of *current* which may flow. It has, however, been observed and verified (according to a well-known mechanical law) that the current that will flow in any given circuit is equal to the electromotive force, divided by the resistance. This is known as "Ohm's Law," from the name of the scientist who first enunciated it, as applied to electricity.

It is usually abbreviated $\frac{E}{R} = C$; and in plain words means that a unit (or a number of units) of E.M.F., divided by a unit (or units) of resistance, will allow a unit (or units) of electricity to pass through the circuit. Taking for example as our *practical* units (see § 16), the Volt, the Ohm, and Ampère, we say that a pressure of 50 volts, working through a resistance of 25 ohms, gives rise to a current of 2 ampères since $\frac{50}{25} = 2$. In working with the battery, it must be borne in mind, that we have always to do with *two* resistances : one, in the battery itself, due to the resistance presented by the exciting fluid itself to the passage of the current; this is called the "internal resistance," and is usually expressed as R; the other, which is *outside* the battery, is due to the resistance presented by anything placed in the outer circuit between the two elements constituting the cell or battery; this may be conducting wires, lamps, electromotors, or in fact anything through which the current has to pass to equalize the difference of potential between the two elements. This, the "external resistance," is generally abbreviated as r. The formula therefore which expresses the relation which *must* exist between the electro-motive force of a given cell, its internal resistance, the external resistance which may be placed in circuit, and the actual current obtainable, is $\dfrac{E}{R+r} = C$; in which E expresses the electro-motive force in volts, R the external resistance in ohms, r the external resistance also in ohms, and C the current in ampères. In words this reads: "The electro-motive force, divided by the internal resistance added to the external resistance, is equal to the current."

§ 20. It will be evident from this, that increasing the size of a given cell will have no influence on the E.M.F. set up, any more than the increase of size of a kettle would have any influence on the temperature at which water contained therein would boil; this temperature, at ordinary pressures, would always be 100° Cent. or 212° Fahr. whatever might be the size of the kettle. But the size of the plates (*ceteris paribus*) would have a great influence on the current flowing. This is due to the fact that the larger the section of a given body, the less resistance it opposes. Hence the fluid between two battery plates 2″ × 1″ would present double the resistance opposed by the fluid between two plates 2″ × 2″. For example, suppose we were using a plate of zinc and a plate of carbon, each exposing 1 square inch of surface, in a cell containing dilute sulphuric acid, and in another precisely similar cell, with the plates at exactly the same distance apart, we used plates exposing 2 square inches of surface. The E.M.F. of the combination we will put down at 1 volt in either case. In the former case the internal resistance we may put down as 1 ohm; in the latter case, owing to the doubled surface, the internal resistance falls to $\frac{1}{2}$ ohm. Now, if we join the two plates outside the cell with a short length of very stout wire (so as to present an external resistance so small as to be practically negligible) we shall get as the resulting current in the two cases, (1) $\frac{1}{1} = 1$ ampère; and (2) $\frac{1}{\frac{1}{2}} = 2$ ampères. It will be perfectly clear from this, that if we have a number of cells, and connect all the zincs together so as to form practically one large zinc, and all the carbons together to form one large carbon, we shall not increase the E.M.F. in the slightest degree, but shall simply be reducing the internal resistance, and consequently increase the current on the short circuit.[1] This mode of coupling cells is known as "parallel connection" or "multiple arc."

§ 21. There are two methods by which we can get an increase of E.M.F.: the first is by choosing as our active element, and acting fluid, bodies which have a greater affinity (greater energy of combination) than the ones previously employed; the second consists in using a number of cells (proportional to the increase in E.M.F. desired), and so coupling these together, that the E.M.F. excited in the one cell shall be added to that excited in the next, and so on. For instance, supposing we were employing as our active element, zinc, as the acting fluid, dilute sulphuric

---

[1] "On the short circuit," is a convenient way of expressing the idea that the circuit externally is so short as to present no appreciable resistance.

acid mixed with chromic acid, and as the passive element, carbon. Such a combination will give an E.M.F. of 2 volts, almost exactly. (If we were using plates exposing about 4 square inches of surface each, at a distance of about ⅜ of an inch, the *internal* resistance in the cell would be about 0·08 of an ohm.) Now, if we couple up a number of precisely similar cells, so that the carbon of the first is connected to the zinc of the second, the carbon of the second to the zinc of the third, and so on for the entire number of cells used, the first zinc and the last carbon being connected to wires (or other conductors) leading to the outer circuit, we shall find that the electromotive force increases with the number of cells used.

§ 22. This mode of coupling up is known as "series connection." In the example just given, if 25 of the said chromic acid cells were coupled together in series, we should have a total electro-motive of $25 \times 2 = 50$ volts; but since the resistance of the cells is also additive, the *current* in ampères on a short circuit remains precisely the same as it would be with a single cell. This is easily seen if we use the formula indicated by Ohm's law. In the case of the single cell we have: $C = 2$ volts, $R = 0·08$ ohm, hence $\frac{2}{0·08} = 25$ ampères. In the case of the 25 cells we have 25 times 2 volts, or 50, but we also get 25 times 0·08, or 2, hence $\frac{50}{2} = 25$ ampères as before. Apparently, there would be no advantage in coupling cells in series rather than in parallel; but a moment's thought will show us that the *external* resistance will largely modify the result. This leads us to the consideration of the best mode of coupling cells.

D

§ 22*a*. Since we can vary the voltage at will, by increasing or diminishing the number of cells placed in *series*, and diminish the internal resistance by placing them in *parallel*, it is evident that we can couple cells in various and suitable

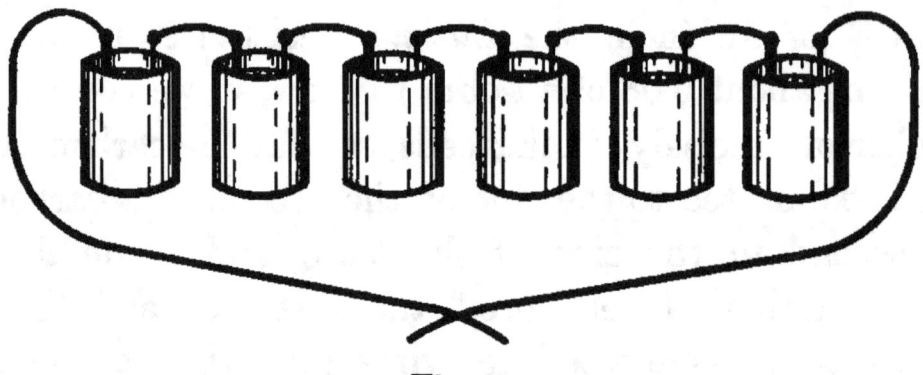

Fig. 6.

manners, to get the desired amount of current through different resistances. As this is a matter of importance in the practical use of batteries, we give four examples of

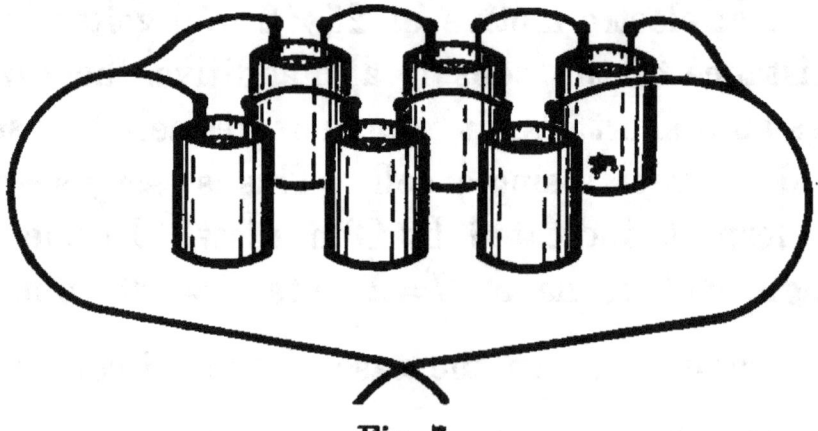

Fig. 7.

the possible modes of arranging six cells, showing how the arrangements will cause the E.M.F. and the internal resistance to vary. In Fig. 6 we represent all six cells joined in series; in Fig. 7 we have two parallel sets of

three in series; our Fig. 8 represents three parallel sets
of two in series; while finally Fig. 9 shows all the six cells

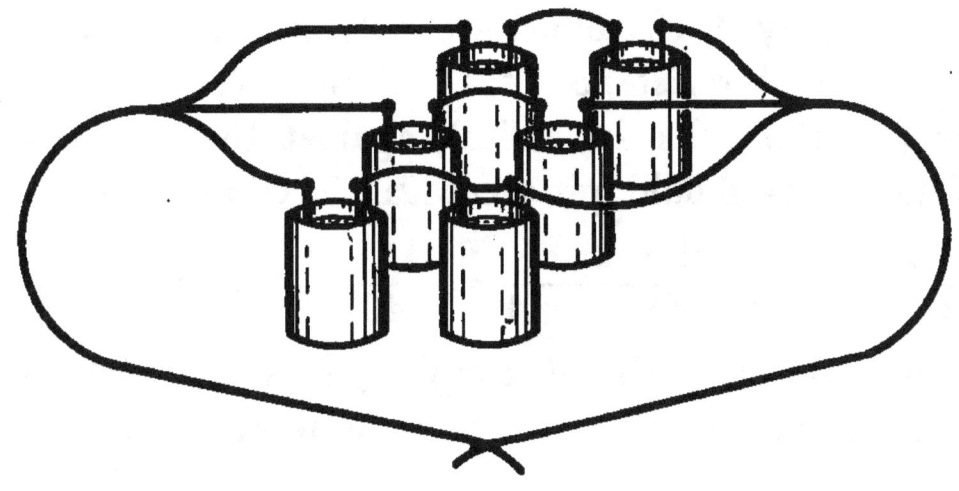

Fig. 8. .

joined in parallel, thus forming practically *one large cell*

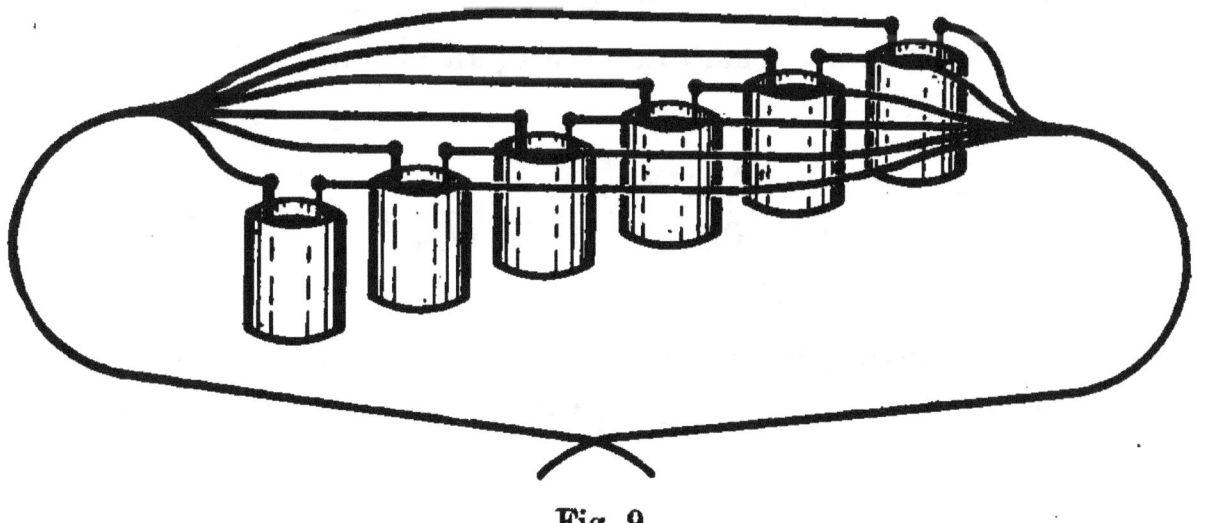

Fig. 9.

exposing six times the superficial area of the elements
that one cell alone exposes. Adhering still to the chromic

acid type of cell which we took as our previous example,[1] having an E.M.F. of two volts, with an internal resistance of 0·08 ohm, we will consider the effects of these four different modes of coupling up: first, with a low external resistance, say of 0·01 ohm; secondly, with a comparatively high external resistance of 11 ohms. Now in the first case (all the cells in series), against the low external resistance, we should get according to Ohm's law—

$$\frac{6\,E}{6\,R + r} = \frac{6 \times 2}{6 \times 0·08 + 0·01} = \frac{12}{0·49} = 24·4 \text{ ampères flowing.}$$

In the second mode of coupling (two parallel sets of three in series), with the same external resistance, we have—

$$\frac{3\,E}{\dfrac{3\,R}{2} + r} = \frac{3 \times 2}{\dfrac{3 \times ·08}{2} + 0·01} = \frac{6}{0·13} = 46 \text{ ampères.}$$

In the third example (three parallel sets of two in series), the same external resistance being in circuit, the following is the result :—

$$\frac{2\,E}{\dfrac{2\,R}{3} + r} = \frac{2 \times 2}{\dfrac{2 \times ·08}{3} + 0·01} = \frac{4}{·064} = 62·5 \text{ ampères.}$$

Lastly, if with the same external resistance of 0·01 ohm we use the six cells in parallel, we get the following equation :—

[1] This is a very convenient cell for experimental work, having a high E.M.F., small internal resistance, and quickly mounted and dismounted. The elements are zinc and graphite plates, the exciting fluid chromic acid three parts, oil of vitriol three parts, water twenty parts.

$$\frac{E}{\frac{R}{6} + r} = \frac{2}{\frac{\cdot 08}{6} + \cdot 01} = \frac{2}{\cdot 023} = 87 \text{ ampères.}$$

From this it is clear, that when the external resistance is lower than the internal more current will flow if the cells are coupled in parallel, than when they are arranged in series. Using exactly the same cells arranged as before, but with the external resistance of 11 ohms, the resulting current, in the four cases, will be as follows :—

(1) $\dfrac{6\,E}{6\,R + r} = \dfrac{6 \times 2}{6 \times \cdot 08 + 11} = \dfrac{12}{11 \cdot 48} = 1 \cdot 04$ ampère.

(2) $\dfrac{3\,E}{\dfrac{3\,R}{2} + r} = \dfrac{3 \times 2}{\dfrac{3 \times \cdot 08}{2} + 11} = \dfrac{6}{11 \cdot 12} = 0 \cdot 53$ ampère (a trifle over ½ ampère).

(3) $\dfrac{2\,E}{\dfrac{2\,R}{3} + r} = \dfrac{2 \times 2}{\dfrac{2 \times \cdot 08}{3} + 11} = \dfrac{4}{11 \cdot 05} = 0 \cdot 36$ ampère (just over ⅓ ampère).

(4) $\dfrac{E}{\dfrac{R}{6} + r} = \dfrac{2}{\dfrac{\cdot 08}{6} + 11} = \dfrac{2}{11 \cdot 013} = 0 \cdot 18$ ampère (about ⅙).

From this we gather that when the external resistance is very *high* as compared to the internal resistance, the largest possible current will be obtained by arranging all the cells in series; on the other hand, if the external resistance be very low, the best results are attained by coupling the cells in *parallel*. Apart from this it may be shown mathematically that the maximum result in any combination of cells, as far as flow of current in ampères is concerned, is obtained when the total internal resistance and the total external resistance are equal. Of course it is not always either convenient or desirable to get this result, but simply to get the largest possible current with the number of cells and with the particular external resistance at hand; in which case, the nearest possible approximation to equality between the two resistances will give the better result. In other words, in arranging a battery so as to get the greatest power for any given work, the cells must be so joined that the number arranged *in series* shall be as nearly as possible the mean between the number of cells, and the number which expresses the proportion between R (the internal resistance) and r (the external resistance). Supposing we have 20 cells, and we find that $r = 6\,R$, or in other words that the proportion of R to r is as 1 to 6, we get as the mean between the number of cells and the proportion number $\dfrac{20 + 1}{2} = 10\frac{1}{2}$, but as we cannot divide 20 into two sets of $10\frac{1}{2}$, we arrange them into the highest nearest possible, which is 10; hence two sets in parallel of 10 cells in series will give the largest possible current under these conditions. Say for example $E = 1$ volt per cell, $R = 1$ ohm per cell, and $r = 6$. Then the total of the two parallel sets of 10 in series is 10 volts, the internal resistance in each set of 10 is 10 ohms, but as these two sets are coupled in parallel, this resistance is divided by 2, hence falls to 5 ohms; consequently we get $\dfrac{E}{R + r} = \dfrac{10}{5 + 6} = \dfrac{10}{11}$ of an ampère, which is the highest possible current obtainable in these circumstances.

§ 23. For this reason it is well to be able to measure

the internal resistance of any given cell. It must be borne in mind that this internal resistance depends on several circumstances. Firstly, the distance of the two elements or plates from each other (since the nearer they stand, the less the resistance). Secondly, the nature of the fluid between the elements. Pure water presents an enormous resistance, saline solutions less, while acid solutions, up to a certain point of acidity (varying with the nature of the acid), conduct fairly well, and consequently present the least resistance. Thirdly, the chemical changes which go on during the action of the cell, changes which generally cause the resistance to rise, in some cases gradually, in others very rapidly. Consequently the internal resistance of any given cell (except perhaps those of the so-called "constant" type) can only be stated for that particular cell, and at the particular time of measurement. There are several methods of ascertaining and measuring the internal resistances of single cells, or of batteries. We will describe the simplest and most useful. The first, and the simplest method, provided the operator has access to a trustworthy voltmeter, and to an ammeter (the resistance of which is known, or has been previously ascertained by means of a Wheatstone's bridge), depends directly on Ohm's law, viz. $\dfrac{E}{R}\,C$; therefore $\dfrac{E}{C} = R$ (see Fig. 10). We begin by measuring the E.M.F. of the cell, by means of a voltmeter; we find, for example, that this amounts to 1·5 volt. We then connect the cell momentarily by means of a tapping key to the ammeter (of which we have previously measured the resistance and found it to be say 0·075 of an ohm). The ammeter indicates 4 ampères. Hence as above $\frac{E}{C} = R$, or

$$R = \frac{1\cdot5}{4} = 0\cdot375 \text{ of an ohm.}$$

Of this, we know 0·075 to be due to the R of the ammeter itself, hence the internal resistance of battery must be 0·375 minus 0·075, or 0·3 of an ohm. The second method consists in arranging a battery or cell so as to be able to couple up in series with any galvanometer without actually connecting up; then

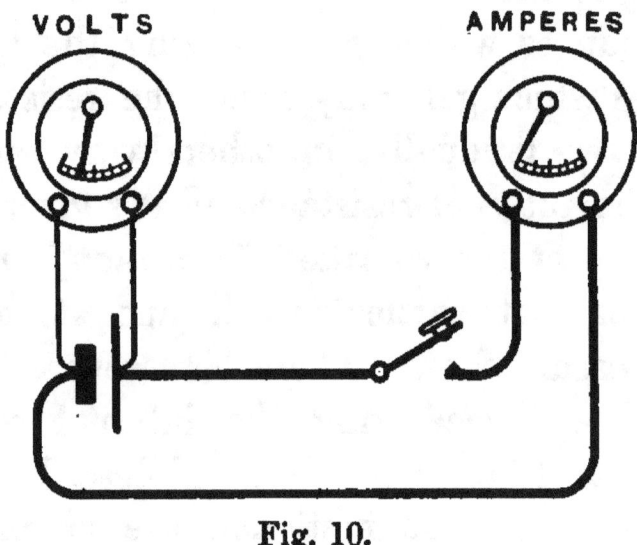

VOLTS   AMPERES

Fig. 10.

taking carefully the resistance of the entire circuit (without the battery) by means of a Wheatstone's bridge. Having ascertained the resistance of the outer circuit, prepare a length of wire having *exactly* the same resistance as the outer circuit, connect this across the poles of the battery *in shunt*, and at the same time couple up to the galvanometer as at first arranged. Note carefully the deflection. Now the current divides itself along two paths; one half goes round the shunt, and one half round the galvanometer circuit. The shunt must now be removed, which causes the entire current to pass round the galvanometer circuit, and this circuit being now broken at any point, resistance is added at this point until the deflection of the galvanometer is brought exactly to what it was before. This added resistance will be equivalent to the internal resistance. Fig. 11 shows the arrangement. *C* is the cell or battery, of which it is desired to measure the internal resistance; *G* is the galvanometer, and *g g g g* the

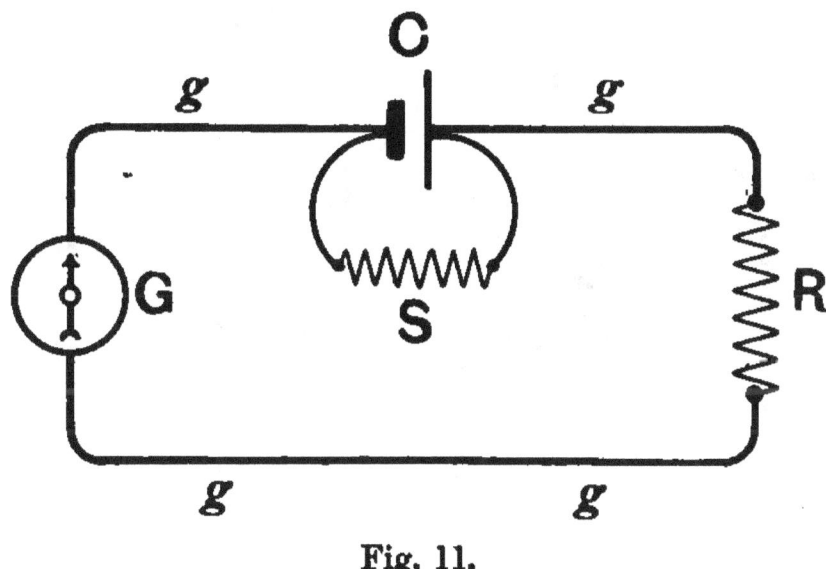

Fig. 11.

connecting wire; *S* is the shunt wire, having the same resistance as *G g g g g* together, *R* is the added resistance, which is employed to bring down the reading to its original point, *after* the shunt *S* has been removed. These two methods are open to the objection that if during the trial the E.M.F. of the battery or cell varies (which it very often does, especially if a heavy current be taken off) the results will not be correct. A third method, which is free from this defect, is the one known as Mance's. It consists in arranging a Wheatstone's bridge and galvanometer in circuit with the cell to be tested, as shown in Fig. 12, with a short-circuit key between B and D. Resistance R must be added in the arm of the bridge, between D and A, until on depressing the key the deflection of the galvanometer does not vary. When this occurs the internal resistance will be equal to R plus $\frac{a}{b}$. If the

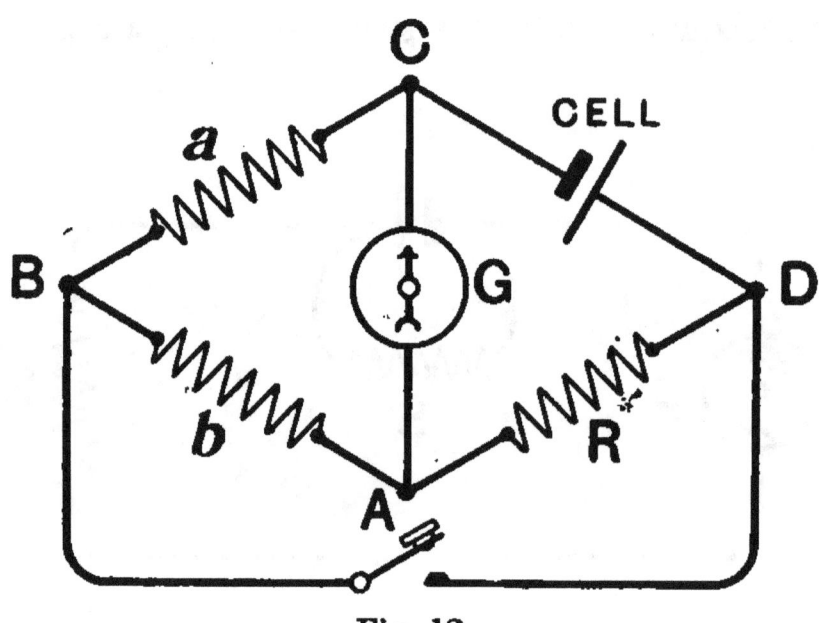

Fig. 12.

resistance employed in the arms a and b be equal, then the internal resistance is equal to R. There are many other methods, varying in delicacy and in accuracy; we shall, however, describe only one other, known as Munro's, which is applicable to every kind of battery. Munro's method, of which our Fig. 13 gives a good idea, consists in arranging a condenser C (the capacity of which may be from $\frac{1}{3}$ to 1 micro-farad) in circuit with the battery B to be measured, a galvanometer G, and a tapping key K_1.

At any point between the battery and the condenser is arranged a shunt S, connected to a second tapping key K_2. To use this arrangement K_1 is pressed down and the deflection (d_1) of the galvanometer noted; still keeping the key K_1 depressed, K_2 is also depressed, and the deflection (d_2) which takes place in the opposite direction is read off; the "internal resistance is equal to the resist-

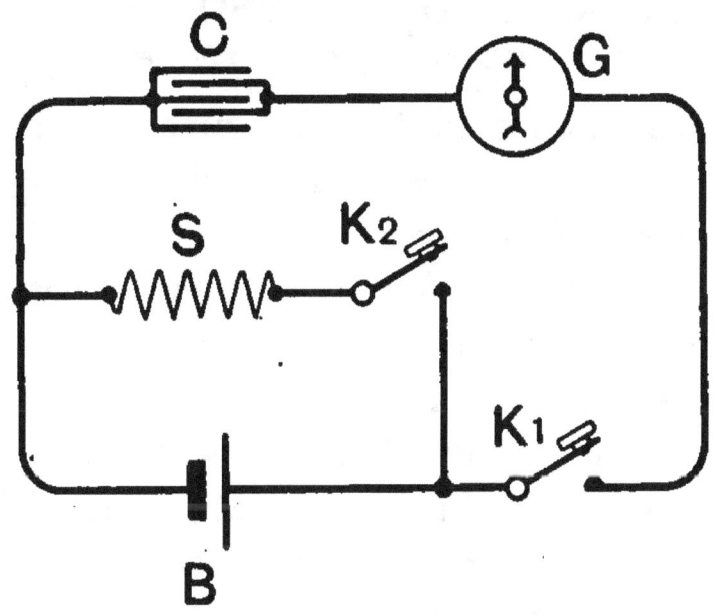

Fig. 13.

ance of the shunt S, multiplied by the deflection d_2 divided by d_1 less d_2," or $R = S \dfrac{d_2}{d_1 - d_2}$.

§ 24. If we plunge a plate or strip of CHEMICALLY PURE zinc in dilute sulphuric acid, this latter has no action on the pure metal; but if a plate of copper, of silver, or of graphite, etc. be immersed in the same acid fluid, and the extremities be allowed to touch, either directly, or by the intervention of a copper wire, or indeed any other conductor, the acid immediately attacks the zinc, dissolving it with the formation of zinc sulphate and the extrication of free hydrogen, in the form of minute bubbles, which are carried along by the current set up, and go towards the copper, etc. plate, where part escape and part adhere to the surface of the copper plate. The chemical action which takes place is represented by the annexed equation :—

$$Zn + H_2SO_4 = ZnSO_4 + H_2.$$

Free zinc and   sulphuric equal zinc sul- and free
acid   to   phate   hydrogen.

It must be remembered, that by virtue of this chemical action, *a current is set up*. But when using the pure metal as above, if the connection between the zinc and the other plate be broken, the chemical action CEASES, and no difference of potential is apparent; no waste of zinc therefore occurs, except only during the time that current is flowing. This of course is of great importance in the economical working of a battery, in which we are practically consuming zinc to obtain energy in the form of electricity, just in the same way as we consume fuel to get energy, in the form of heat, in our coal fires. Now if instead of *pure* zinc we employ the ordinary commercial article, which contains a small amount of other metals, specially iron, cadmium, traces of copper, etc., we find that chemical action goes on, when the zinc is immersed in the dilute acid, whether a plate of a dissimilar metal be in the solution or not, or whether this second plate be disconnected or otherwise. Hence, whether the cell be giving current or no, the zinc is being consumed, and in the latter case being simply wasted. The theory generally accepted, to account for this "local action" which takes place when ordinary commercial zinc is immersed in dilute acid, is that the minute particles of dissimilar metals present in the impure zinc (iron, copper, cadmium, etc.) act themselves like a set of negative plates (copper, silver, platinum or graphite) in contact with the zinc, thus favouring the development of the electric current, which flows *locally*, that is to say, from the zinc particles to the iron, etc. particles on the plate, by this means permitting the chemical action to take place, whether the other metal plate be connected to the zinc one or not. If this theory be correct, the name "local action," which is used to denote this peculiar behaviour of ordinary commercial zinc, expresses very fairly at once the cause, and the position, of this result. When commercial zinc is used as the "positive" or current starting element, in conjunction with a negative (or current receiving) element, either of copper, silver, platinum or graphite, it is found that by amalgamating the zinc plate with *mercury*, we confer upon it the peculiar immunity to the action of dilute acid which is the characteristic of pure zinc.

§ 25. There are several modes by which zinc can be amalgamated. One very good way is to clean the surface of the metal of adhering grease, by the aid of ordinary washing soda and warm water, then, after having rinsed it, to rub over the surface with a pledget of tow which has been dipped in dilute sulphuric acid (say water 12 parts, acid 1 part, by weight), and while rubbing to pour a small drop of mercury on the surface, to which the mercury immediately adheres, and can be made to spread itself all over, by continued rubbing. If this is nicely done, the surface of the zinc becomes entirely amalgamated, and shines as resplendently as burnished silver. A second plan, which gives a very well and evenly amalgamated surface, but in which the mercury does not generally penetrate so deeply, and consequently the good effects are more fugitive, consists in making a solution of mercury either in dilute nitric or sulphuric acids,[1] and applying the one or the other of these solutions to the surface of the zinc plate or rod, with a rag or pledget of tow as before, directly after the said surface has been freed from grease by washing with soda and rinsing with water. Whichever method be adopted, care must be taken not to use so much mercury as to penetrate right through the zinc, as in this case the zinc becomes extremely brittle and fragile. The effect of amalgamating the zinc is very remarkable: it not only confers upon it immunity from the action of acids (with the exception of nitric acid) when the circuit is broken, but the zinc amalgam has a higher E.M.F. than the unamalgamated metal. It is noteworthy, however, that if a saline solution be used as the excitant, amalgamation has little or no *protective* effect.

§ 26. It is difficult to explain with any degree of certainty *how* the mercury acts to prevent local action. Many hold the opinion that the mercury dissolves only the *pure* zinc, and forms a superficial coating of this pure zinc amalgam on the surface, which therefore acts

---

[1] These salts can be purchased ready made, under the names of mercury nitrate and mercury sulphate respectively. They must be dissolved in water for use.

practically like a plate of pure zinc. Against this view may be cited the fact that *mercury* is itself *another metal*, and not only so, but a metal that is strongly negative to zinc. Others lean to the opinion that the effect of amalgamation is largely mechanical; and that the immunity conferred on zinc by amalgamation is due to the

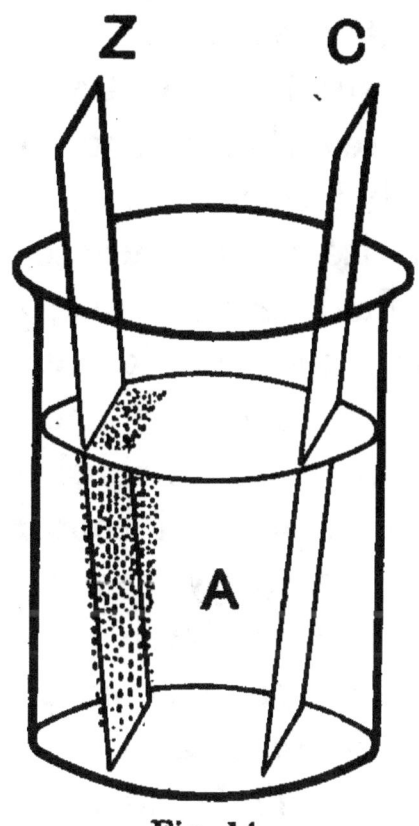

Fig. 14.

fact that a very smooth mirror-like surface is produced on the zinc, which on being plunged in the dilute acid becomes immediately coated with a layer of adherent hydrogen gas (see § 24) that protects it from further action, unless indeed the circuit be closed by connecting through any conductor the two elements in the cell, when of course this adherent hydrogen is carried forward to the negative element, where it appears, and is partly liberated in the form of small bubbles.

§ 27. The student would do well, in order to impress this latter result on his mind, to perform a simple experiment which will render this evident. Let *A*, Fig. 14, represent a glass jar, containing dilute sulphuric acid (1 acid, 12 water). Into this is plunged a strip of ordinary zinc *Z*, and at a little distance from it, not touching the zinc at any point, a similar strip of copper, silver, or carbon, *C*. It will be noticed that minute bubbles of hydrogen gas are immediately formed on and about the zinc, which bubbles will soon rise to the surface, and break away at the surface of the fluid near the zinc, giving it the appearance of boiling (see the equation at § 24). If now the plates be inclined towards one another so as to touch above, or if connection be established between the two plates by means of a piece of copper wire, or any other good conductor of electricity, a very peculiar effect will be observed. Few or no bubbles of hydrogen will be extricated at the zinc plate or strip; but these bubbles will now make their appearance at the strip or plate *C* (see Fig. 15). Many will be liberated there, and at the commencement the apparent boiling (due to the extrication of these bubbles) will proceed rapidly, but as the action goes on, in a less lively manner. If by means of an ammeter, or any form of galvanometer capable of measuring the amount of current given off, we note the strength of the current at the beginning and during the progress of the experiment, we shall find that the current rapidly falls off. It will also be found, however, that by removing the strip C, and wiping its surface, or even by agitating the fluid in the vicinity of this strip (which can easily be done by "stirring" the fluid with the plate or strip C), the current strength rises almost to its original value. Not quite, because the strength of the acid is continuously weakened by its becoming saturated with zinc. From this we learn two important facts: first, that

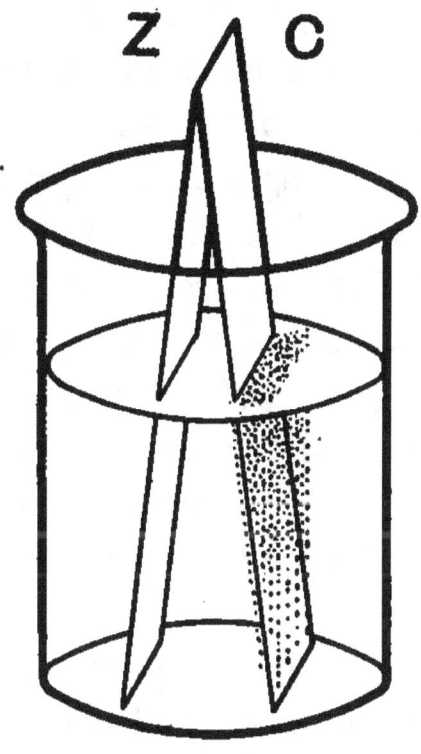

Fig. 15.

during the electrolytic action of a cell, the liberated hydrogen is carried by or along with the current from the positive to the negative plate; secondly, that although much of this hydrogen *is* liberated in the form of bubbles on the surface of the liquid in the proximity of the negative or "collecting" plate, yet sooner or later (dependent largely on the nature of the surface of this negative plate) sufficient hydrogen collects there to coat the surface with a layer or stratum of gas, which being itself electro-positive, like zinc, rapidly levels up the difference of potential originally existing between the positive or "energy supplying" plate, and the negative or "energy receiving" plate, by covering the surface of the latter with a stratum which is at least nearly as positive as the zinc itself, thus practically converting the negative plate into a positive one. This peculiar effect is known by the very erroneous name of "polarization," and is a great source of loss of constancy in nearly all cells and batteries of the "single fluid" type.

§ 28. Polarization, when used in this sense, may therefore be defined as "the alteration of the condition of the surface of the negative or collecting plate by the adherence of a film of hydrogen gas." This is the sense in which the word is generally used; but it must be quite evident that the deposition or even the adherence of any electro-positive body on the negative element of a cell, must produce a similar effect, and we shall see later on, that by taking advantage of the power we have of varying the electrical condition of one or of both the elements in a cell, it has become possible to construct batteries in which current is evolved during the restoration of the original state of things, from the polarized to the non-polarized state.

§ 29. In most cases, so far from being any assistance in the production of current, this phenomena of polarization is a serious drawback. Not only do the E.M.F. and current rapidly decrease in consequence, but in the case

of ordinary commercial zinc, consumption of zinc and acid
go on to very much the same degree. Several methods
have been therefore suggested to get rid of this adherent
hydrogen, which is the cause of polarization. Agitation
of the fluid, favouring the escape of the gas bubbles;
heating the fluid, to the same end; drawing air through
the liquid by means of an aspirator; mechanically brushing
off the gas from time to time, have been suggested and
used. Other means have been adopted by different
workers to compass the same intention, such as roughening
the surface of the negative plate (which facilitates the
extrication of the gas bubbles), either by scratching the
surface, by the electro-deposition of platinum in a powdery
condition, or again by making the negative element in the
form of a gauze made out of copper or silver wire woven
into a mesh; and lastly, the addition of some substance
in the cell, which has the power of combining with the
liberated hydrogen. These bodes are therefore called
depolarizers; and according to requirements, or to the
nature of the cell, may either be used in the ordinary
form of cell, in which both the positive and negative
element stand in the same fluid mixed with the depolarizer, or may be used in a separate porous cell, in which
the negative element stands, while the porous cell in its
turn stands in an outer vessel, containing the positive
plate along with the exciting fluid.

CHAPTER IV

SUBDIVISIONS OF CELLS INTO SINGLE FLUID AND DOUBLE FLUID PRIMARY AND SECONDARY

§ 30. BASING ourselves on the fact that some cells are constructed without any depolarizer being used, or if such exist, without any provision being made for the depolarizer being kept separate from the exciting fluid proper, we can conveniently subdivide galvanic or voltaic batteries into two great classes, viz. "single fluid" and "double fluid" cells. In the single fluid cell, of which we give a typical illustration in Fig. 16, we have at A a containing vessel, to which can be given any convenient form, containing the exciting fluid B, that may or may not have in admixture with it some substance capable of seizing on the liberated hydrogen, and thus act as a depolarizer as well as an excitant; standing in this fluid we have two plates, or rods, or sheets, Z and C, of which the one marked Z consists of some metal (or other substance) which is readily acted on chemically by the exciting fluid; and the other, C, of some metal or other good conductor of electricity, on which the exciting fluid has little or no action. This latter plate C serves to collect and transmit the electrical disturbance set up by the chemical action of the fluid on the plate, rod, or sheet Z. Because the chemical action starts at the element Z, and the E.M.F. takes its origin at this spot, the flow of the current *in* the cell is *from* this element Z *to* the element C; hence it is usual to term the element

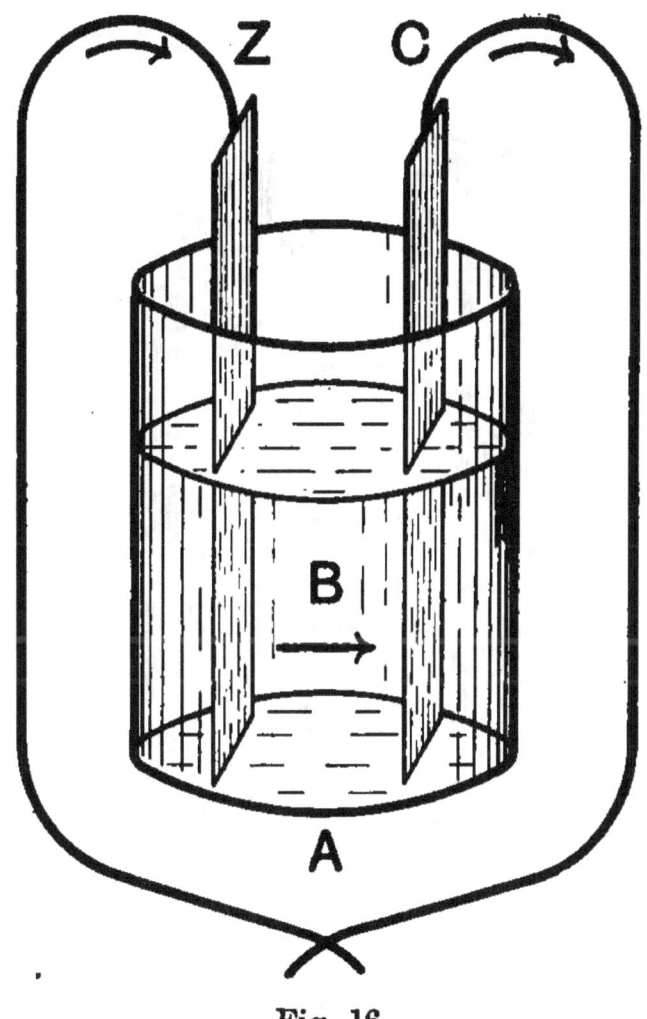

Fig. 16.

Z "positive," as having the higher level of energy, or "potential," while the element C, which is receiving and passing on this current due to the difference in potential, is usually called negative; but this only holds good when we regard the condition of the two elements *in* the cell.

It will be abundantly evident, after a little consideration, that when the circuit is completed between the two elements, by a wire or any other conductor, the part played by the two elements *outside* the cell is reversed. Here the plate *C* sends the current along the wire to the

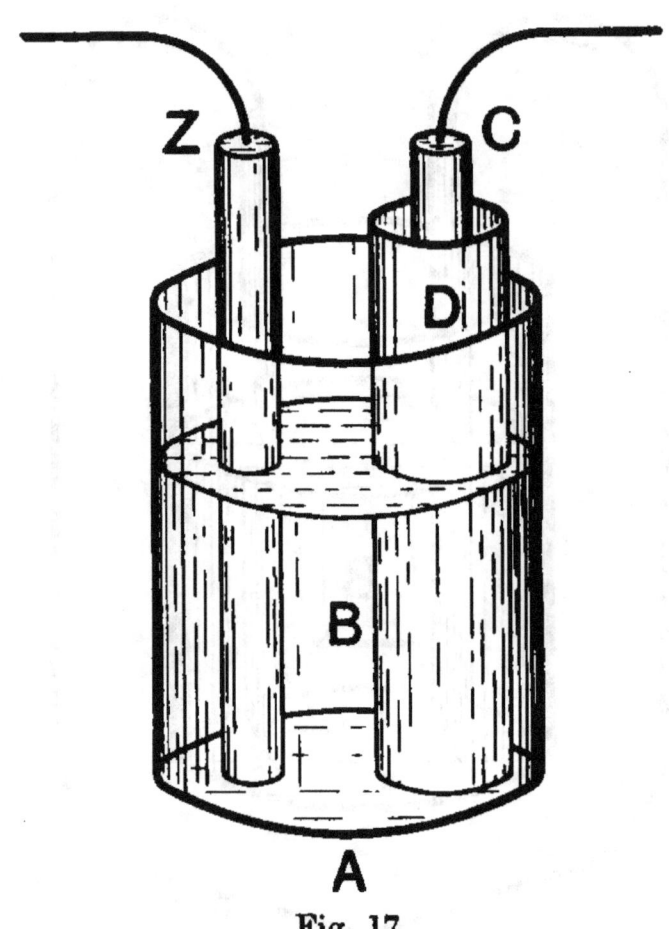

Fig. 17.

plate *Z*, to restore the disturbed equilibrium; hence the element *C* is *positive* OUTSIDE the cell, while the element *Z* is *negative*.

In the double fluid cell, of which Fig. 17 conveys a good general idea, we have as before a containing vessel *A*, in which is the exciting fluid *B*, and the active or positive element *Z*. Instead of allowing the negative element *C* to stand directly in the same exciting fluid, it is separated from it by means of a porous pot *D*, which contains not only the element *C*, but also the solution of some body *E* which has the power of seizing upon the hydrogen and converting it into water, or by some other means preventing it from adhering to the surface of the negative element *C*. The intelligent reader will at once perceive, that the relative positions of *Z* and *C* may be varied to suit varying conditions; and that provided the depolarizing solution always accompany *C*, and the excitant be in contact with *Z*, it is indifferent whether *Z* or *C* be placed in the porous pot (or other porous partition). The pot or other septum *D*, employed to keep the excitant and the depolarizer from mixing too freely with each other, must be porous; at least sufficiently so as to be so far permeated by the two fluids as to allow the entire division to be fairly conducting, otherwise the resistance presented to the passage of the current would be so great as to render the combination practically useless. As we shall see later on, not only is current passing along, but chemical action *must* take place in the fluids permeating the pores of the porous pot or septum. It will be needless to remark, that the direction of flow of the current is the same in the double fluid as in the single fluid cell; and that the same rules hold good with reference to the opposite conditions of the elements *in* and *out* of the fluid; *Z* being *positive* to *C in* the fluid, while *C* is positive to *Z outside* the fluid. As it is usual to call the extremities of the two elements (as well as the wires connected to them) which are out of the excitant by the name *poles*, so it is customary to designate *Z* as the *positive* element, and the wire attached to it the *negative pole;* while for a similar reason, *C* is termed the *negative* element, and its wire, the *positive pole* of the cell or battery.

§ 31. Shortly after the discovery of the pile by Volta, Mr. Nicholson showed the decomposition of water under the influence of voltaic current; and a careful repetition of his experiments, with modifications by Sir Humphry Davy, brought to light an apparently extraordinary fact, namely, that when a compound body is decomposed [1] by current electricity, its constituents are not only separated from each other, and carried, the one to the negative pole and the other to the positive pole of the battery effecting this decomposition, but, furthermore, that during this apparent transport or "migration" of the constituents, the said constituents lose their peculiar chemical activity; or perhaps it will be more correct to say, that while being transported by the current, their usual chemical characteristics are masked. As precisely the same thing takes place IN the cell or battery itself, it will be well, in order to obtain an intelligent idea of this action, that we should recapitulate Sir H. Davy's experiments, and point out the bearing which they have upon the modern accepted theory of the "migration of the ions." If we arrange three wine-glasses *A*, *B*, *C*, side by side as in Fig. 18, and join them together by means of a few strands of moistened darning-cotton *D*, *D*, having previously nearly filled the glasses

---

[1] A body capable of being decomposed by the passage of a current is termed an "electrolyte."

with a solution of sodium sulphate, coloured with either
blue litmus or infusion of red cabbage, we shall find on
connecting them up to four or five cells of any ordinary
battery, that decomposition takes place, as is rendered
evident by the change of colour that takes place in the
bluish vegetable infusion, which becomes red in the glass
A, wherein dips the *positive* pole of the battery, and
green in the glass *C*, at which the negative pole enters.

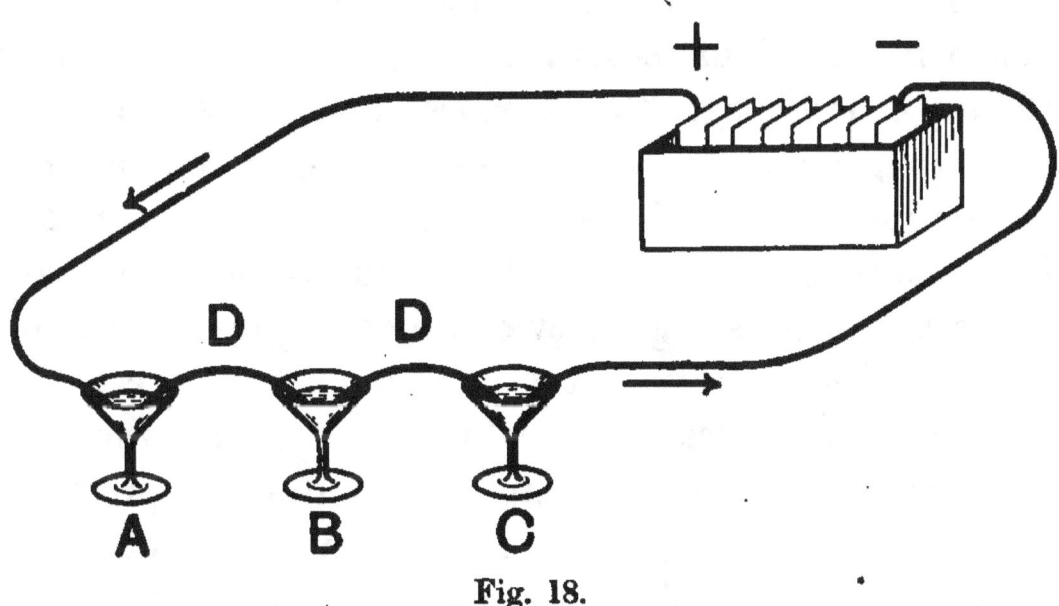

Fig. 18.

So far the result is not surprising; the peculiarity lies in
the fact, that even if the decomposing action of the current
be kept up until the whole of the sodium sulphate in the
three glasses has been decomposed, glass *A* finally containing all the sulphuric acid, and glass *C* all the soda, so
that the sulphuric acid must have traversed from *C* through
B to reach *A*, and, *per contra*, soda must have traversed
through *B* on its way to *C*, yet the blue vegetable colour
in the glass *B*, *at no time* changes its colour.   This is

sufficiently striking in itself; but it might be partially explained by supposing that in traversing the glass *B*, in equivalent proportions, the soda and the sulphuric acid neutralized each other, were it not for the fact that even if we vary the experiment by putting the coloured sodium sulphate solution in one of the outer glasses only (say for example in *C* alone), while the other two contain a watery solution of the blue colouring matter, precisely the same results are obtained; and the liberated sulphuric acid traverses the central glass *B* without effecting the slightest change in its colour, sensitive though it may be, under ordinary conditions, to the presence of the merest trace of acid. In glass *A* (as before) the litmus, or red cabbage infusion, immediately shows, by its reddening, the presence of free sulphuric acid, while the simultaneous liberation of soda in the glass *C* gives evidence of its presence by the greening of the solution in this glass. From these two experiments (the results of which hold good, whatever be the nature of the salts employed in solution, provided only that the resulting liberated constituents do not form compounds too insoluble to remain in solution or partial solution) we learn two facts. (*a*) Compound bodies which are soluble, or which can be reduced to a fluid state by fusion or otherwise, are split up into their component parts, under the influence of the current; the electro-negative elements going to the positive pole, or "electrode," of the battery immersed in the solution, and the electro-positive constituent going to the negative pole or electrode. (The term "electrode" means simply a "way" or passage, and is derived from the Greek word "οδος.")

(*b*) That during this decomposition, no matter where-abouts in the circuit these soluble compound bodies[1] be situated, provided they be in fluid connection with each other, an *apparent* transport of the constituents takes place, the electro-negative element or elements *seemingly* being carried along with, or by the current, to the positive electrode, while the electro-positive element is in like manner driven to the negative electrode.

§ 32. It is evident, however, that no actual transfer of the two constituents of the electrolyte could be actually taking place, if by *transfer* we are to understand the actual passage of the liberated constituent, from the pole at which it was liberated, to the opposite pole. For this reason Grotthus so far back as 1805 formulated an hypothesis which, in conjunction with slight additions suggested by Clausius, forms the accepted theory of to-day, and which satisfies the requirements of observed facts. Briefly stated, the theory is this. All chemical elements have a certain amount of energy : those which possess *more* being *electro-positive* to those which are endowed with less. A chemical compound is the result of the union of an electro-positive element with an electro-negative. Hence all known elements may be arranged in a tabular form, so as to give an idea of their relative electrical bias ; for instance, we may say that the metal potassium is electro-positive, while the gas fluorine is electro-negative to every other known element.[2] When a compound is in solution (or in

---

[1] Usually termed "electrolytes" in this connection, meaning bodies which can be *split up* by the *electric* current.

[2] We give a table of the more important elements, arranged in the order of their electrical tendencies, at p. 64.

a fused condition) we are to suppose that the molecules of this compound are always in movement, gliding about amongst one another; and not only so, but that the constituent atoms of these molecules are also in movement, continually separating and recombining to form similar groups, these movements taking place in *all possible directions throughout the liquid.* (This is quite in conformity with the modern kinetic hypothesis of the constitution of liquids.) If there be no electro-motive force, and no electrified body present, to upset the balance between the positive and negative constituents of the molecules, it is evident that there can be no " inducement," so to speak, for the atom to move more in any one direction than in any other. Let us suppose, however, that we now introduce into the electrolyte two oppositely electrified bodies (not in contact with each other). Now, in obedience to the well-known law that similarly charged bodies repel, while dissimilarly charged bodies attract one another, there will be a tendency for these molecules to arrange themselves in certain definite positions. In other words, under the electric strain thus produced, the molecules will tend to arrange themselves with their *positive* constituent atoms looking towards the *negative* body, and their *negative* constituent atoms facing the positive body, thus forming a *polarized* chain. We might also say, that the motion, which before the introduction of the two oppositely charged bodies was uniform in every direction, becomes suddenly *controlled* as to direction. If, now, the electric strain be kept up (either by means of a separate battery, or by chemical action between the bodies introduced and the electrolyte itself), this atomic motion, being controlled in

direction, takes a more definite form, and results in an
exchange taking place between the positive atom of the
molecule nearest the positive body and the positive atom
of the molecule next in sequence, which in its turn displaces the positive atom of the adjacent molecule, and so
on all along the chain of polarized molecules, very much
in the same way as couples exchange partners in a dance;
until on arriving at the negative body, the last displaced
positive atom finds itself set free, and adheres to this
negative body, be it an electrode or an actual negative
plate.   It will thus be seen, that at no time is the dissociated
atom actually free, but simply passing hands (to continue
the simile of the couples in a ball-room), till in its turn it
finds itself at the opposite end of the chain.   It must be
remembered, that while this effect is going on *in one
direction* with the positive atoms, one precisely similar,
but opposite in direction, is going on with regard to the
negative atoms, and the negative body immersed in the
electrolyte.   In order to render this perfectly clear, we
give an illustration of a simple case, in which a zinc plate
is used as the electro-positive body (which we shall distinguish by the chemical symbol $\overset{+}{Zn}$), hydrochloric acid
being employed as the electrolyte, and a carbon plate as
the electro-negative body.   As the molecule of hydrochloric acid is built up of one atom of hydrogen (which
is electro-positive) and one atom of chlorine (electro-
negative), we can represent it by $\overset{+}{H}\ \overset{-}{Cl}$.   The carbon,
being electro-negative to zinc, we will distinguish by the

chemical symbol $\overline{C}$. In our Fig. 19 the first line of
hydrochloric acid symbols illustrates the "unarranged"
or "directionless" condition of the molecules of the
electrolyte, *before* they are subjected to any electrical

strain by the immersion of the plates $\overset{+}{Zn}$ and $\overset{-}{C}$. Line

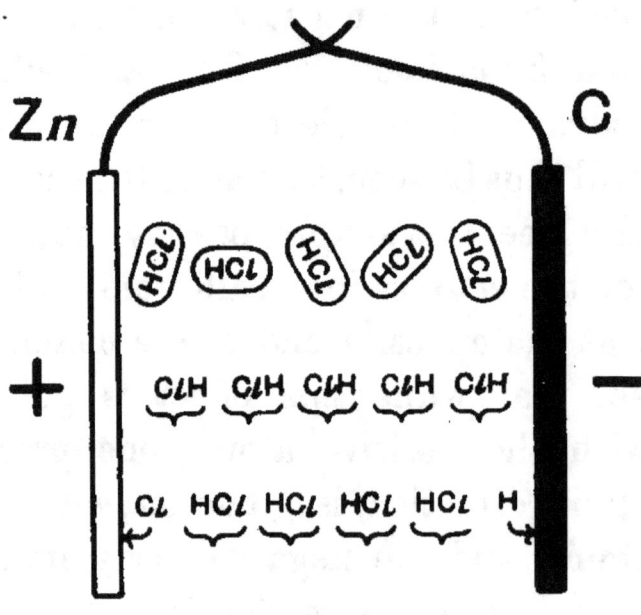

Fig. 19.

two shows the "arranged" or "polarized" condition of the
molecules when subjected to the strain due to the introduction of these plates, but with no current passing; and
lastly, line three gives an idea of the "change of partners"
which takes place along the polarized chain when the
current is flowing. According then to this theory, at no
part of this chain is any one atom of hydrogen actually
set free from a chlorine atom until it arrives, by a kind
of "changing partner" process, at the negative element

$\overset{-}{C}$, where, as there is no chlorine atom with which it can enter into union, it is liberated in the form of gas. So in like manner, but in the opposite direction, each atom of chlorine, under the repulsion effect of the negative element $\overset{-}{C}$, and the attractive effect of the positive element $\overset{+}{Zn}$, is impelled to change place with the chlorine atom in the next farther molecule of the chain, and so on all along the line, atom by atom, from molecule to molecule, the chlorine changes place, until the positive element $\overset{+}{Zn}$ is reached, where it is finally liberated. This explains *why* the constituents of a compound undergoing electrolysis do not give evidence of their peculiar chemical characteristics except when they reach the vicinity of the negative and positive electrodes respectively, since they are at no time really *free* until they reach the extremity of the chain. Subjoined is a list of the more important elements, beginning with those which are the most highly electro-positive, and terminating with the one which displays the highest electro-negative bias. Each element in this list is electro-positive to the ones below it, and electro-negative to the ones above it :—

TABLE OF CHEMICAL ELEMENTS

ARRANGED ACCORDING TO THEIR ELECTRICAL CONDITION

Positive end

Hydrogen
Potassium
Sodium
Magnesium
Aluminium
Zinc
Cadmium
Tin
Iron
Nickel }
Cobalt }
Lead
Copper
Silver
Palladium
Gold
Carbon (graphite)
Platinum
Iridium
Rhodium
Nitrogen
Antimony
Tellurium
Selenium
Sulphur
Iodine
Bromine
Chlorine
Oxygen }
Fluorine }

Negative end

CHAPTER V

ON SECONDARY CELLS

§ 33. THE knowledge of the facts discussed in our last chapter will enable us to form an intelligent idea of the principles on which the action of those useful pieces of apparatus, erroneously called "storage cells," "accumulators," or, with more accuracy, "secondary batteries," depends. If it be attempted to pass a current of electricity through *pure* water, this fluid presents so high a resistance as to render the electrolytic effect almost nil; but if the water contain a trace of saline matter, or better, of sulphuric acid, say $\frac{1}{30000}$ part, it becomes a fairly good conductor, so that the water is freely electrolyzed, and split up into its constituent elements, hydrogen and oxygen. The following equation shows the final result of this splitting up of the water under the strain of the electric current :—

$$H_2 O = \quad H_2 \quad + O$$

Water equal to hydrogen and oxygen.

What this equation does not show, is that the cause of this decomposition is the attraction that the negative electrode or "cathode" has for the positive hydrogen conjoined to the attraction displayed by the negative

F

oxygen for the positive electrode or "anode." Bearing in mind what was stated in Chapter II., that all elements are endowed with a specific amount of energy, which energy is released when two or more combine together, it must be evident that this act of decomposition is necessarily effected by a restoration of that amount of energy to the constituent elements which is essential to their existence in the free state. We may liken the decomposing effect of the current to the mechanical strain exerted in separating two helical springs from each other, when nearly relaxed and in contact at the nearer extremities. It will be easily perceived that when the strain is removed, the springs will tend to return to their original position, again giving up the energy imparted during extension. Both the decomposing effect of the current, and the tendency to return to the original state, can be well and easily shown by an arrangement similar to that illustrated at Fig. 20, in which *A* and *B* are two glass tubes, in the upper extremities of which are sealed two fairly stout platinum wires, reaching to the bottom or open end of the tubes, and terminating in binding screws, mercury cups, or other means of connecting to a battery, as shown at *P* and *N*. These two tubes can be made to fit accurately into the two necks of a Woolf's or similar double-necked bottle *C*. The tubes are first removed, the bottle filled with slightly acidulated water, after which the tubes are inserted and the bottle turned upside down, so as to fill *them* with water. The bottle is now placed in its upright position, and connection made with a battery *Q* having an E.M.F. of at least 2·5 volt, which is about the amount of energy required to tear the atom of oxygen from the two atoms of hydrogen.   Under these circumstances the water in the tubes *A* and *B,* as also in the bottle *C,* is decomposed, the hydrogen collecting round the platinum wire in the tube *B,* and rising in the same, gradually displacing the water therein.   In like manner oxygen gas is liberated

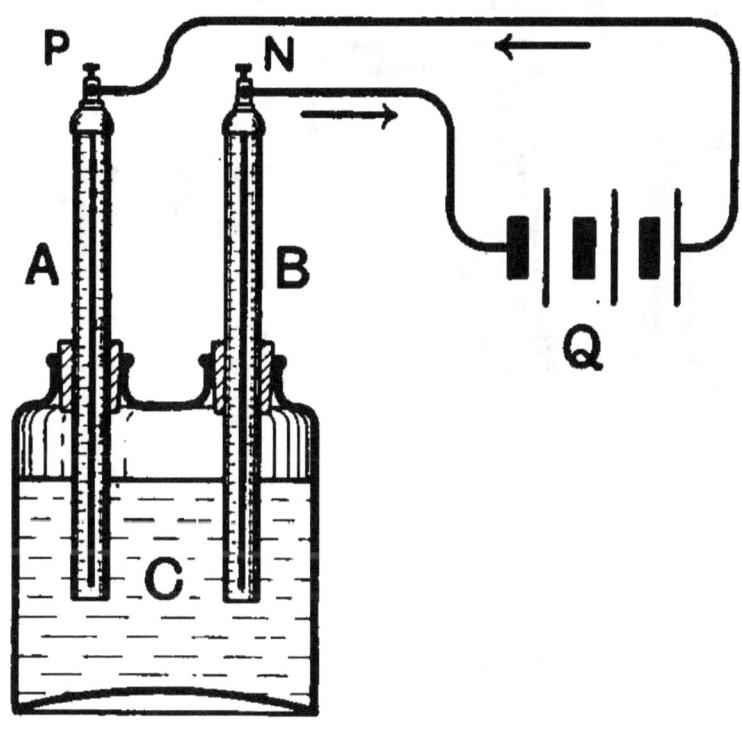

Fig. 20.

in the tube *A.*   During the progress of this experiment it will be noticed that *in the same time* the current liberates twice the bulk of hydrogen as it does oxygen, according to the formula for water H_2O (see Fig. 21).   So much for the electrolytic or electro-decomposition effect of the current.   If, after the action has been going on some little time, so as to secure the liberation of a fair amount of the two gases, we connect, by means of two wires, the platinums P and N (see Fig. 22) through a galvanometer, G, or any other appliance for recognizing the presence of electricity flowing, we shall find that a current is being set up, and traverses the circuit from N to P; or, in other words, in the opposite direction to that in which the

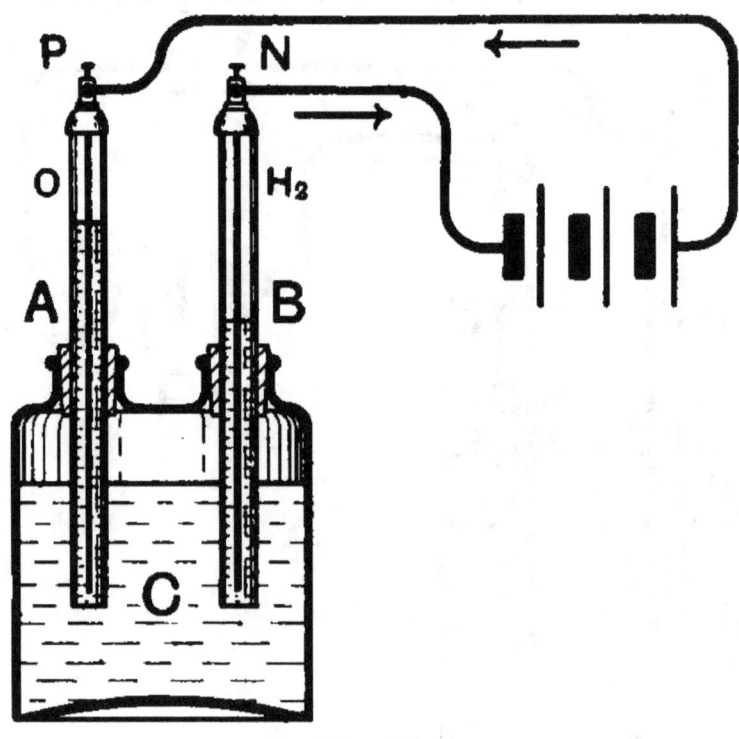

Fig. 21.

current was originally sent by the battery; while, at the same time, the amount of the two gases diminishes, until after some time the gases themselves disappear, and the current ceases. The cause of these two effects is not far to seek : relieved from the strain put upon them by the current forced through by the battery, the two gases slowly recombine to form water, giving up at the same time *in the form of electricity*, that energy which (having been imparted to them by the battery current) had enabled them to exist as separate gases. So true is this, that it is quite possible to get the same effect, no matter how the two gases have been prepared: a similar result accruing, if instead of preparing the gases electrolytically from water, we fill the two tubes, the one with oxygen prepared by heating chlorate of potash, and the other with

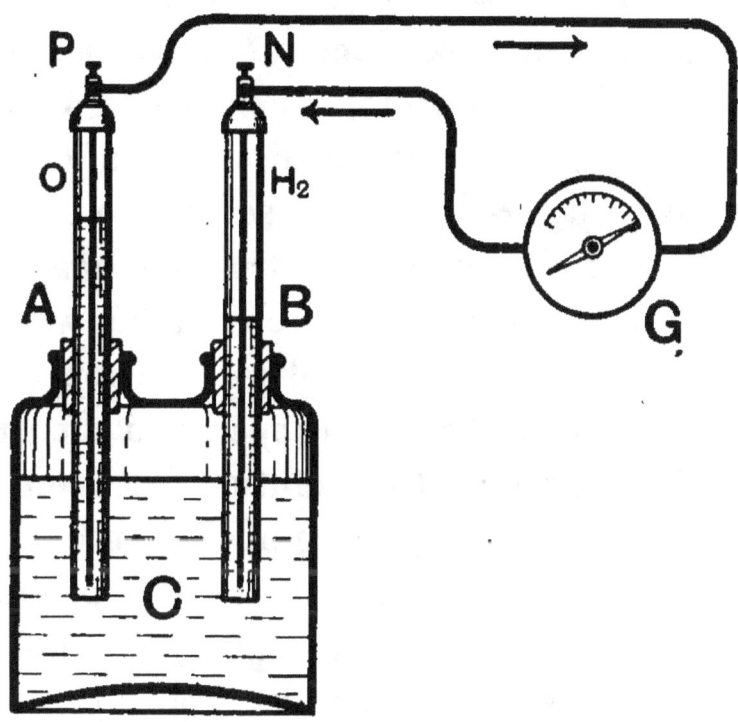

Fig. 22.

hydrogen obtained by passing steam over red-hot iron borings contained in a gun-barrel.

§ 34. This "gas-battery" gives us a very clear idea of what a secondary battery really is. Generalizing on this, we may define the secondary cell as being one in which the contained fluid is by itself incapable of acting upon the other elements of the cell, and thus liberating energy; but that when energy (in the form of electricity) is applied

to them, the fluid is split up into its electro-negative and
electro-positive constituents, which go to adhere respectively to the elements of opposite signs, rendering *them*,
pro tem., negative and positive to each other; and that
these liberated constituents, having energy re-imparted to
them, can and do react on each other as soon as electrical
communication is established between them outside the
cell, by means of any conductor, *but not before*. Because
they are capable of returning *as electricity* a large proportion of the energy thus imparted to them, they are sometimes called *storage cells*. It is needless to say, that many
other bodies which can be electrolyzed, may be used to
produce effects similar to those obtainable with water and
two platinums. Thus if two plates of copper be plunged
into a solution of sulphate of copper, and a current passed
for some time through the solution between the plates,
the solution is gradually decomposed, the copper plate
connected to the negative pole of the primary battery
becoming coated with metallic copper, while the liberated
sulphuric acid attacks the positive copper plate, re-forming
sulphate of copper at its surface. The *chemical* reaction
that takes place may be expressed, as far as the liberation
of the copper and of the sulphuric acid, *and also* the decomposition of some of the water holding the copper sulphate
in solution are concerned, by the following equation :—

$$H_2 O + Cu SO_4 = O + SO_4 + Cu + H_2$$

Water   Copper sulphate   Oxygen Sulphuric Copper Hydrogen
radical

At the negative plate, then, we have a coating of metallic
copper, with adherent hydrogen; at the positive plate we
have adherent oxygen, while the copper plate is partially

dissolved by the liberated SO_4, with the formation of fresh $Cu\ SO_4$ (copper sulphate). As another instance, we may take the result obtained by electrolyzing water acidulated with sulphuric acid, two sheets of lead being employed as the electrodes. Let *P* and *N* (Fig. 23) represent two sheets of lead, plunged into dilute sulphuric acid contained in the jar *J*. On connecting these up to a battery

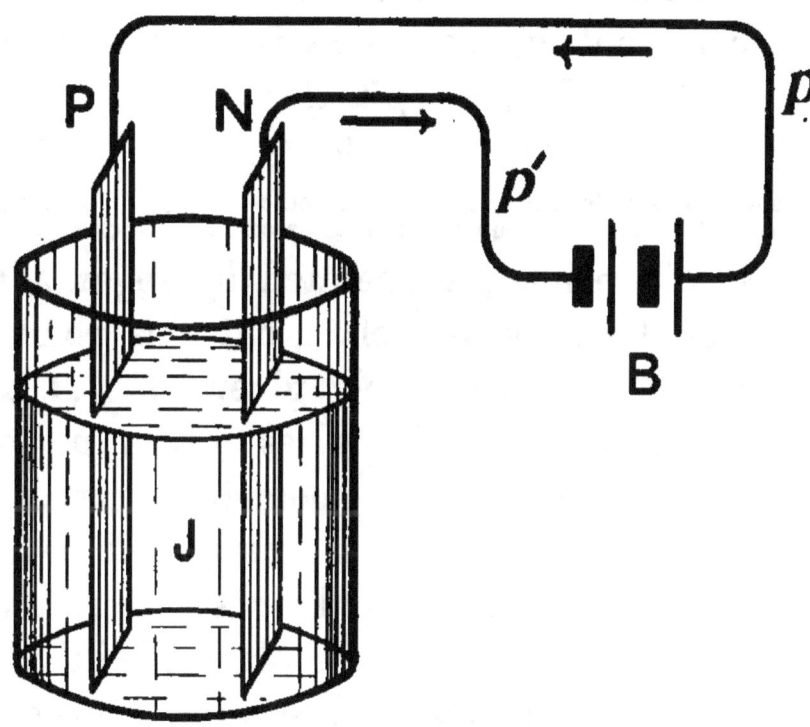

Fig. 23.

B, capable of giving current at not less than 2·5 volt-pressure, it will soon be noticed that the plate *P*, which is attached to the wire proceeding from the positive pole of the battery, changes tint, assuming a rich chocolate brown, or puce colour, while the other plate *N* alters but little, or at most becomes darker. At the beginning but few bubbles are evolved at *N*, but after some time an extrication of minute bubbles of hydrogen gas will be noticed, and finally the fluid becomes milky-looking, owing to the diffusion of a quantity of very minute bubbles of this gas in the mass of the liquid.[1] During this reaction, the water alone appears to be decomposed; the sulphuric acid acting only by rendering the water a good conductor, and thus facilitating the passage of the electric current. The chemical reactions which take place under the stress of the electric current are shown in the annexed equation, in which Pb stands for lead (*plumbum*):—

$$Pb \quad + \quad 2\,H_2O \quad = \quad Pb\,O_2 \quad + \quad 2\,H_2$$

| 1 atom of lead | and | 2 molecules of water | are equal to | 1 molecule of lead peroxide | and | 2 molecules of hydrogen. |

Up to a certain point the liberated hydrogen adheres to the surface of the *N* leaden plate, but this point is soon reached, and gas is then freely given off. If now the battery be removed, and the wires *p* and *p'* be joined together (better through a galvanometer or other current indicator) as shown at Fig. 24, a current will be found to flow from *P* to *N* precisely as in the gas-battery previously described. At the same time the plate *P* will be seen to lose its puce colour, gradually becoming greyer and yet more grey, until it finally assumes its original lead colour. Here we have a visual proof that the source of the current is the recombination of the elements, hydrogen and oxygen, which had been forcibly separated by the energy imparted to them. The chemical reaction which takes place during this recombination is expressed below :—

---

[1] As this experiment is very easy of performance, and gives a good practical insight into the theory of the accumulator, the student is strongly recommended to try it himself.

$$\overset{-}{Pb\ O_2} \quad + \quad \overset{+}{2\ H_2} \quad = \quad 2\ H_2O \quad + \quad Pb$$

1 molecule of lead peroxide and 2 molecules of hydrogen equal to 2 molecules of water and 1 atom of lead.

We might multiply examples to almost any extent, since nearly every substance that can be electrolyzed by being subjected to a current when lying between two plates of

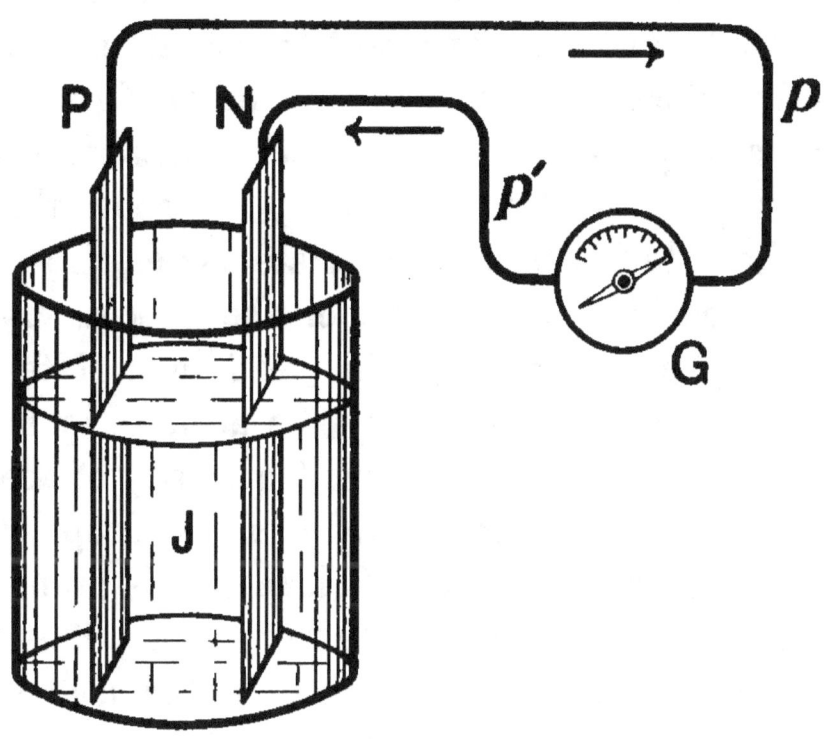

Fig. 24.

like metal (or other conductor) will give somewhat similar results. But it is needless to go further into the subject here, since when we come to treat of the accumulator we shall notice all those forms of secondary cells which have proved themselves to be of any practical use.

§ 35. As in the case of the gas-battery described in § 33, so in this form of secondary cell, it is not necessary that the energy supplied to the two plates, *P* and *N*, should be derived directly from a battery; as a matter of fact it is perfectly feasible to cause two lead plates to give current when immersed in acidulated water, provided we previously coat the surface of the one with finely divided lead, and that of the other with lead peroxide, both of which may have been prepared by chemical means. It must be, however, noted, that in obedience to the well-known law that elements in their "nascent"[1] condition are much more active chemically than at any other time, a pair of plates thus prepared do not give such energetic results as those freshly coated by electrolysis. To show this result without the aid of current to separate the electro-negative and electro-positive constituent of the "secondary" cell, we need only precipitate metallic lead in a state of fine division from one of its salts, say the acetate, by means of metallic zinc. To this end a small quantity of acetate of lead is dissolved in water, to which a few drops of nitric acid have been added. A rod of clean zinc is plunged into the solution, when in virtue of the superior affinity of the zinc for the acid, it is able to displace the lead from its combination, giving up to it at the same time that amount of energy which is necessary to its existence in a free state. The following expresses the chemical interchanges which take place:—

$$Pb\ Ac\ +\ Zn\ =\ Zn\ Ac\ +\ Pb$$
Lead acetate and zinc equal zinc acetate and lead.

Here Ac stands for the radical of acetic acid.

The precipitated lead thus prepared is to be collected

---

[1] An element is said to be in a "nascent" (freshly born) state, when it has *just been* separated from any of its combinations.

and pressed over the surface of one leaden plate, to which it will adhere if the surface have previously been scored.

In like manner the lead peroxide may be prepared by making ordinary red-lead into a stiff paste with sulphuric, spreading this mixture over the surface of the other lead plate by means of a wooden spatula, allowing to dry, and then plunging the pasted plate into chloride of lime-water, which converts the red-lead mixture into peroxide. These two plates thus chemically coated with finely divided metallic lead (electro-positive) on the one hand, and with lead peroxide (electro-negative) on the other, will be found to give (when plunged into dilute sulphuric acid, and the elements joined by a wire) a powerful current, precisely similar in nature and direction to that obtained in the case described in the latter portion of § 34, which current will continue flowing until the negative plate has parted with all its surplus oxygen to the positive plate.

§ 36. In studying these interesting effects, it is well to bear in mind, that by whatever means we are enabled to impart the positive and negative conditions to the opposing plates, be they of gold, platinum, copper, lead, etc., and whatever be the cycle of changes which the fluid in the cell has to undergo before it parts with one or more of its electro-positive constituents on the one hand, and gives up its electro-negative on the other, the final and all-important result (as far as the "charging" effect is concerned) is the accumulation of the electro-negative constituent on the positive plate simultaneously with the heaping up of the electro-positive constituent on the negative plate ; and that with regard to the efficiency of the

combination in giving off current when the primary source
of energy has been removed, and the circuit completed
between the two plates, it will depend on the facility
with which the resulting electro-positive and electro-
negative constituents, thus forcibly separated, can recombine with each other.   It does not necessarily require that
the two separated ions should be able to recombine immediately to re-form directly the original compound, as in
the case of the gas-battery; the recombination may even
take place in several distinct steps, and therefore be much
more complex in character than that illustrated by the
equations we have selected as examples, as in the case of
the modern accumulator; but naturally the more direct
and simple the steps required to effect the recombination
of the liberated ions, the more efficient *ceteris paribus* will
the secondary cell be as a restorer of the energy originally
supplied.

CHAPTER VI

TABULATION OF THE DIFFERENT CELLS, ETC.

§ 37. HAVING made ourselves conversant with the requisites to produce a good voltaic cell, we can take up the thread which we broke at the end of Chapter I., and follow the gradual evolution of the battery as originally devised by Volta, till we reach the more compact and efficient forms in use at the present day.

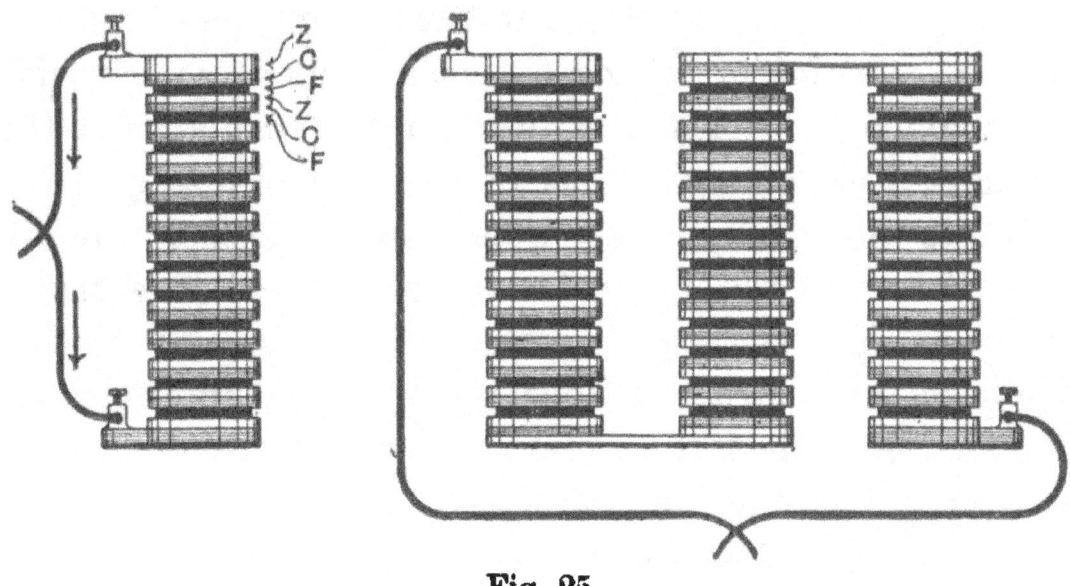

Fig. 25.

Volta's first arrangement (1800) consisted, as shown at Fig. 25, in a number of circular zinc and copper plates, about

$2\frac{1}{2}''$ diameter, $\frac{1}{8}''$ thick, each pair being soldered together, back to back. These pairs were piled one over the other, with all the zinc surfaces looking in one direction, and all the copper ones in the opposite. Between each pair of plates were placed discs of thick flannel or cloth of less diameter than the plates (about $2''$), previously soaked in dilute sulphuric acid. In our illustration Z represents the zinc, and C the copper portion of each circular plate, while F shows the position of the disc of moistened flannel or cloth. When the number of alternations exceeded more than 18 or 20, in order to prevent the weight of the superincumbent mass forcing much of the acidulated water out of the flannel discs, it was usual to build the plates up in two or more columns (as figured to the right in our illustration), connected alternately at the top and the bottom by one or more metallic bars J, and with wires $E\,E$, or vessels of acidulated water, to serve as electrodes. The great point in setting up such a pile, to work with the maximum efficiency, is to avoid the possibility of the acidulated water oozing out of one flannel disc over the plate to the next; which would create a short circuit. The plates are the better for being soldered together in pairs, as there is then no likelihood of the acid getting between the plates, and thus setting up counter electromotive force; but they may however be separate, and merely laid one on the other. In this case the greatest care must be taken to admit no acid between the pairs. When the pile is built up in several columns, care must be taken that each succeeding column begins with the contrary metal to that of the preceding one, otherwise the currents from the two columns would oppose and annul one another. Volta's pile is now only used experimentally for the purpose of demonstration, as it polarizes very rapidly; the flannels retain but little acid, so that it is soon spent. Besides this, it is troublesome to set up.

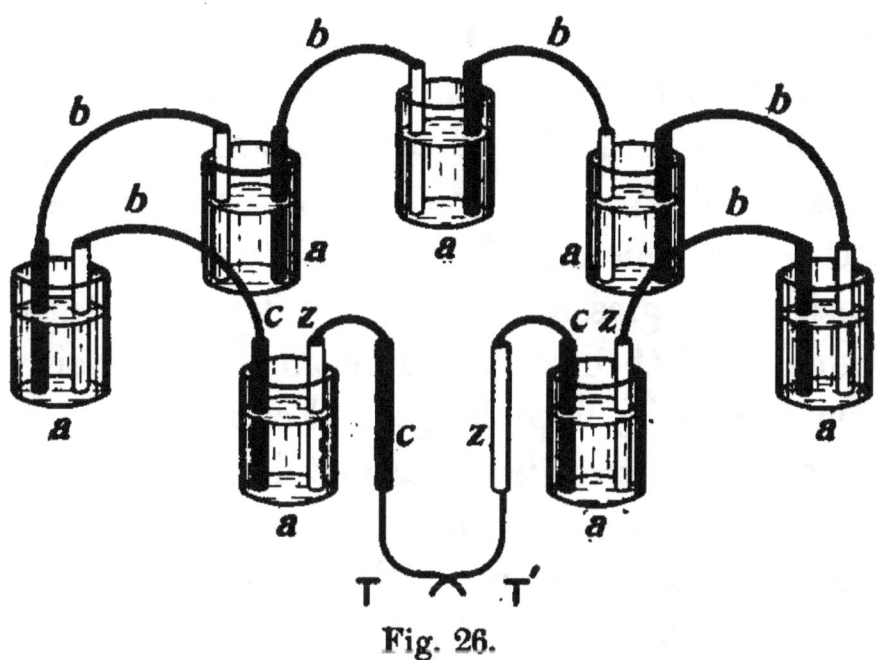

Fig. 26.

§ 38. The next improvement was the substitution of cups, jars, or other containing vessels of insulating material, to hold the copper and zinc elements and also the acidulated water, instead of depending on the flannel to retain the acid. To this combination Volta gave the name, " Corona di Tazze " (French, " Couronne de tasses "), " crown of cups," as he originally arranged them in a circle, for the sake of connecting the pairs together, and of bringing the opposite extremities within convenient working distance of each other. The " Couronne de tasses " forms the subject of our Fig. 26, in which *a a a a a a a* are the glasses containing acidulated water, *z c* the zinc and copper plates, *b b*, etc., being the metallic junctions or solderings of the plates, *T* and *T'* representing the terminal wires or electrodes, by means of which the circuit could be closed either directly between the wires, or indirectly through any conductor or partial conductor, along which it was desired to pass the current. A glance at this arrangement will show that it is much more convenient than the "pile," firstly, because there is but little difficulty in ensuring that no leakage of acid shall take place between cell and cell; secondly, because as the amount of acidulated water which can be contained in the glasses is much greater than that which the flannel or cloth discs could be made to retain, so the duration of activity is prolonged. Following out his idea that the contact of two dissimilar metals was essential to the liberation of electricity, Volta placed a zinc and a copper plate *outside* each terminal cell. This of course was an erroneous conception, and the battery works quite as well if wires only be used as the electrodes *T T'*.

§ 39. Great as was the improvement in point of convenience in setting up, and of duration of action effected by this arrangement, it was felt advisable to give a more compact form to the battery, and to this end was devised the " trough battery," of which we give an illustration at Fig. 27. In this form a trough *T* is made of baked mahogany, with glass partitions; or more conveniently, of glazed stoneware or Wedgwood. Each such trough was usually fitted up with ten or twelve partitions or cells. The plates *P* intended to dip in these cells were connected in pairs, on a strip of baked wood, the zinc of one cell being connected to the copper of the next, and so on in series. Owing to the fact that the plates and the fluid are independent of one another, the former can be let down into the latter when required for use, and can be readily cleaned or replaced when worn or injured, without disturbing the fluid; and the latter may in like manner be removed and changed with the utmost facility. This construction admits of the plates being brought much nearer to one another (an advantage, because the nearer the

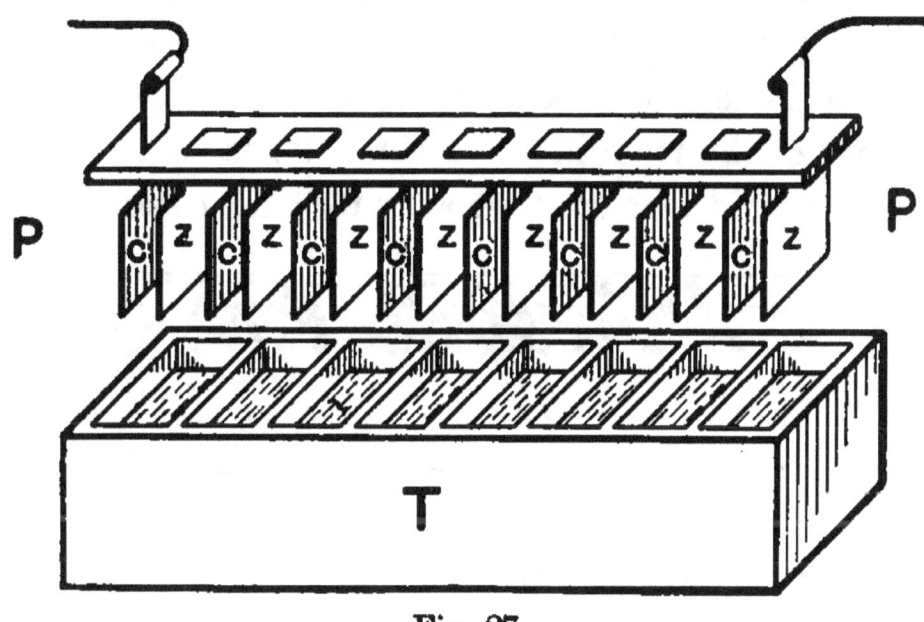

Fig. 27.

plates the less the internal resistance, hence the larger the current). A number of these troughs may be connected up in series, by joining up the terminal plates of the adjoining troughs, care of course being taken, as in the case of the pile (§ 37, latter half), to preserve throughout the series the same order of alternation in the plates, by connecting the zinc end of one battery to the copper end of the next. The voltaic battery which belonged to the Royal Institution, and with which Sir Humphry Davy

made the masterly experiments that led to the discovery
of potassium, sodium, etc., was constructed on this plan,
and consisted in 200 separate troughs, each trough containing ten pairs of plates, and each plate having a superficial area of 32 square inches.   The total number of pairs
was therefore 2000, and the total superficial area 128,000
square inches.

§ 40. Mr. Cruickshanks (1801) modified the trough
battery, by causing the copper and zinc pairs (previously
united by soldering back to back) to form themselves the
partitions of the trough.   For this purpose a suitable box
was constructed of baked wood, with the desired number

Fig. 28.

of transverse grooves in the insides and across the bottom.
Into these grooves were cemented, with pitch or any resinous
compound, the soldered pairs of copper and zinc, care being
taken of course to arrange so that all the zinc surfaces
should look in one direction, and all the copper surfaces in
the other.   The battery was charged by filling the cells
either with a saline solution (salt and water), or with dilute
sulphuric acid, and the galvanic circuit was completed by
bringing the two wires proceeding from the terminal plates
of the battery into contact with one another.   Fig. 28
gives an idea of the general appearance of the Cruickshanks battery, of which Fig. 29 is a sectional view, and

will tend to elucidate the principles of its action. Troughs
of this description (apart from the rapidity with which

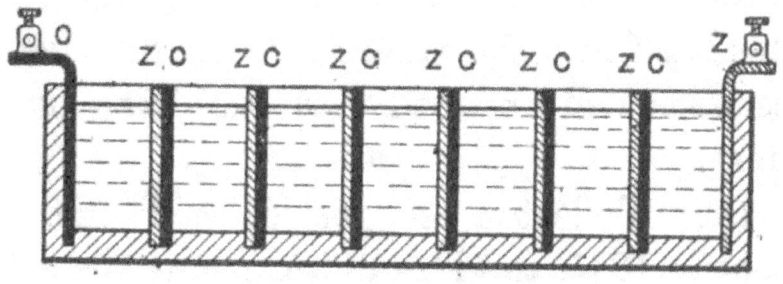

Fig. 29.

their action falls off, owing to polarization, see § 28) are
exceedingly liable to get out of order, from the action of

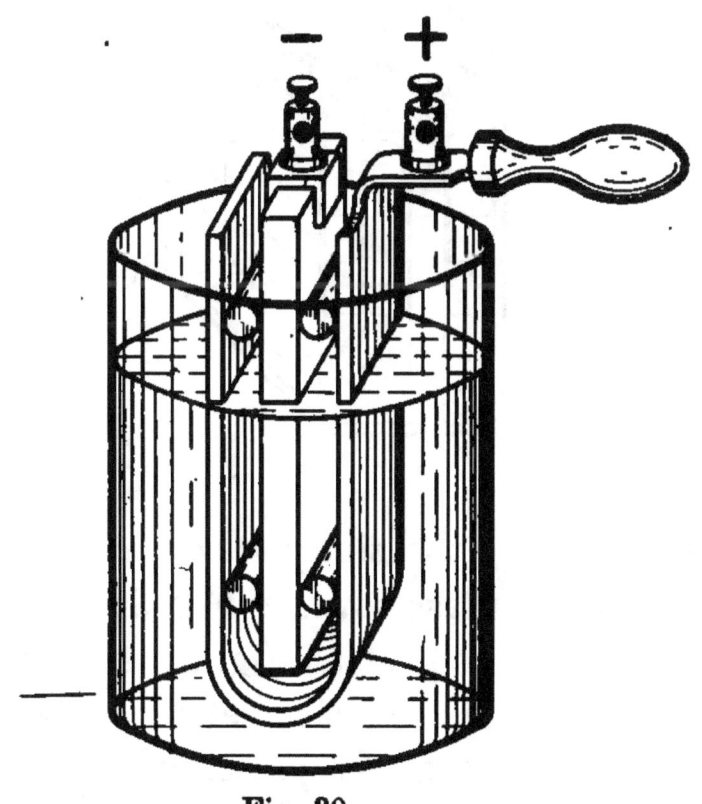

Fig. 30.

the liquid on the wood, which it tends to warp; the plates
require to be fixed in the grooves with some kind of

bituminous or resinous cement in order to render them water-tight; but this cement is apt to crack, owing to the warpage of the wood and other causes, when the liquid insinuates itself into the fissures, thus destroying the insulation between cell and cell, and impairing the power of the battery.

§ 41. Dr. Wollaston suggested, and Mr. Children carried out in 1815, the idea of increasing the efficiency of

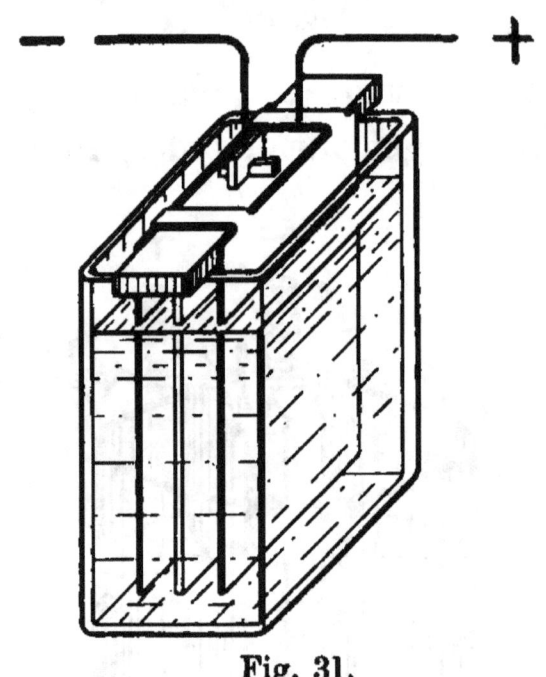

Fig. 31.

the cell, by placing the zinc plate *between* two copper plates connected together, as shown in Fig. 30. In Mr. Children's battery, of which we give at Fig. 31 an illustration of one cell only, each plate was six feet long, by two feet eight inches wide, so that it presented 32 square feet of surface. Owing to the large extent of negative surface exposed, and the proximity of the zinc to the copper plates, the internal resistance of this battery was very low, so that for a short time, and with low resistance in the outer circuit, it gave a very large current.

§ 42. An ingenious modification of this idea was carried out in the battery constructed by Mr. Hart, of Glasgow (see *Edinburgh Journal of Science*, iv. 19), which required no other containing vessel for the exciting fluid than one of the metals which themselves form the battery. This he accomplished by converting the double copper plates into

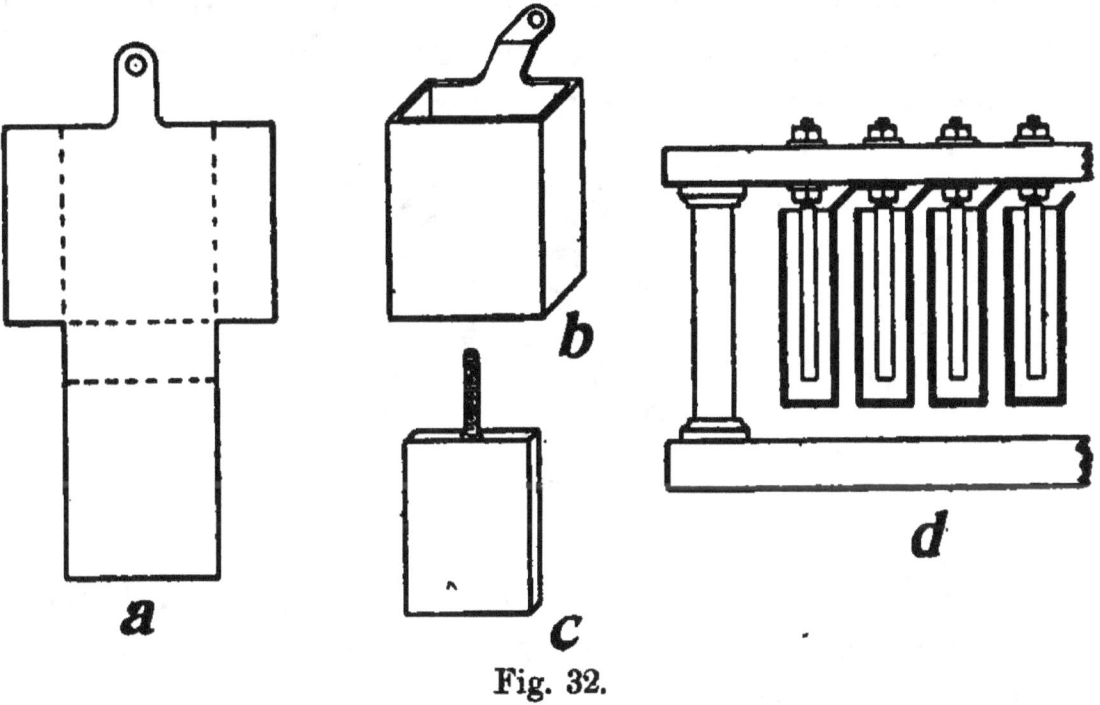

Fig. 32.

cells, with bottoms and sides, so as to enable them to hold the acid solution into which the zinc plates were immersed. Fig. 32 shows the mode of construction. A sheet of copper is cut of the shape illustrated at *a*, then folded as indicated by the dotted lines so as to form a rectangular box, the edges being folded and soldered so as to make the box water-tight. This is shown at *b*. At *c* we have the arrangement by which the zincs are changed to the cross line, which carries the cells. This consists in a screwed brass rod and a couple of nuts. Finally at *d* we illustrate sectionally a portion of the battery fitted up, the lug of the first copper cell being clamped by the nut under the next zinc plate, and so on all along the line of cells. When the battery was to be used, the transverse bar carrying the cells was lifted off the frame, and the cells dipped into a lead-lined wooden trough containing the dilute acid. Such an arrangement was found to give a larger current than any other battery then known having an equal number of zinc plates of the same size.

Fig. 33.

§ 43. Dr. Hare, Professor of Chemistry in the University of Philadelphia (see *Silliman's Journal*, iii. 105), in his battery, the "Calorimotor," took advantage of the increased current to be obtained by exposing a large negative or "collecting" surface, and also by keeping the elements in close proximity to each other, without unduly diminishing the bulk of exciting fluid. This he effected by arranging the copper and zinc elements, as shown at Fig 33, in horizontal section, where the thick line *Z* represents the zinc, and the thinner line *C* the copper. It consisted of sheets of zinc and copper formed into spiral coils, so as to encircle each other, separated by a space only a quarter of an inch in width. The zinc sheets were 9″ × 6″; the copper 14″ × 6″; more of the latter metal being required, as in every coil it was made to commence *within* the zinc, and completely to surround it on the outside. Each complete

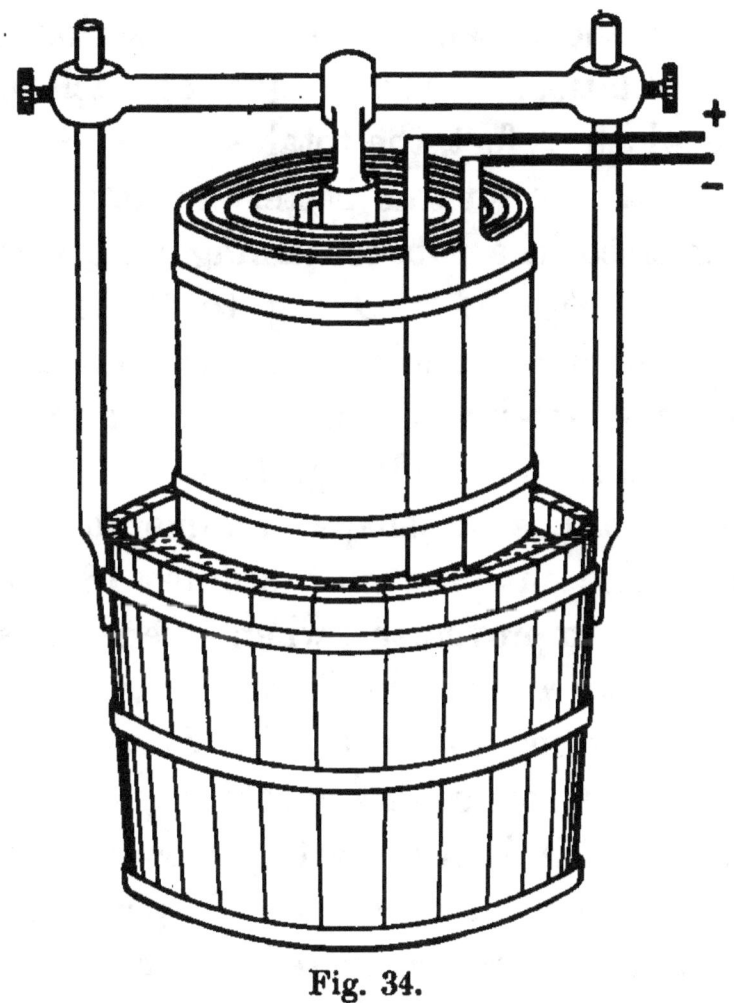

Fig. 34.

combination was about 2½″ in diameter, the total number amounting to 80. By means of a lever they could all be let down at the moment into as many glass jars, 2¾″ inside diameter, 8″ in height, placed so as to receive them, and containing the acid fluid intended to act upon the zinc.

§ 44. A battery very similar to this was the one constructed by Mr. Pepys for the London Institution. It consisted of two plates only, the one of zinc and the other of copper, coiled round a cylinder of wood, and separated from each other by two or more strands of horse-hair rope rolled in between the plates. (This is a fairly good insulator, and at same time is not seriously affected by the acid solution.) The length of each plate was 60 feet, the width 2 feet, the total surface exposed being 400 square feet. To set this battery in action, the whole coil was immersed in a tub containing acid of the proper strength. (See Fig. 34.) Owing to the large current flowing on the short circuit, and the consequent great heating effects obtainable, these two latter forms of batteries were known as " Deflagrators."

§ 45. Modifications in shape innumerable were suggested and carried out between the dates 1816 and 1821, without any real improvement, when Offershaus suggested the use of closely-wound spirals of the two metals, with a view of diminishing the internal resistance of the cell ; but the necessity of so frequently replacing the zinc spirals rendered this form of battery most inconvenient in practical use, without any great gain in point of efficiency.

§ 46. In 1828 Kemp discovered the peculiar action of the amalgamation of zinc, in preventing the action of dilute acid on zinc even if not chemically pure (see § 25), and Sturgeon, in 1830, put this discovery to practical use in his " amalgamated zinc " battery.

§ 47. About this time Münch constructed a trough battery with no partitions at all between the elements, but this entailed great loss of energy, owing to the short circuits which occur between the elements; Faraday, in 1835, to some extent remedied this defect by separating the pairs of elements with heavily-varnished paper. The first real improvement over the plain zinc-copper in acid cell was due to Dr. Alfred Smee, who noticed that the hydrogen gas liberated at the negative plate was evolved from it much more readily, hence polarization took place much less rapidly if the surface of this plate were roughened instead of being quite smooth; and the means he

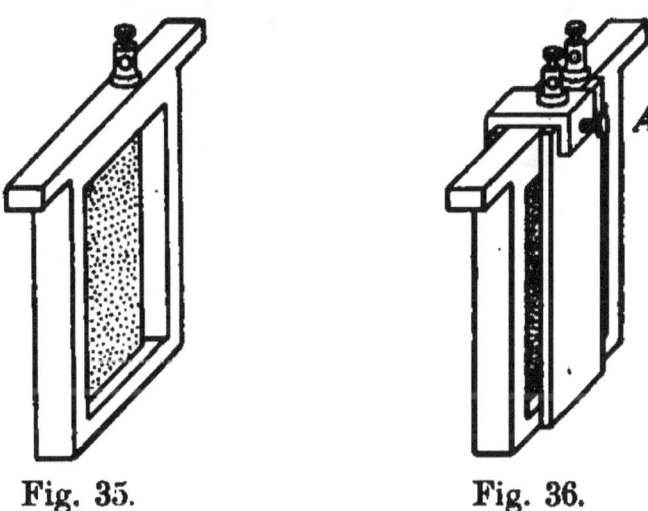

Fig. 35.   Fig. 36.

found most efficient was that of coating the thin silver sheet or sheets with finely-divided platinum, by immersing the silver sheet in a solution of platinic chloride, and using the silver plate as a cathode, when on passing a battery current through the solution the platinum is electrolytically deposited as a finely-divided black powder. The excitant usually employed in the Smee battery is sulphuric acid diluted with water, strength about one part by weight of strong sulphuric acid, sp. g. 1·840, to twelve parts of water. With a view to economizing the silver, this is used in very thin sheets, and a sheet of this platinized silver being clipped in the central bezel of a light wooden frame (like a boy's small slate), this frame is sandwiched between

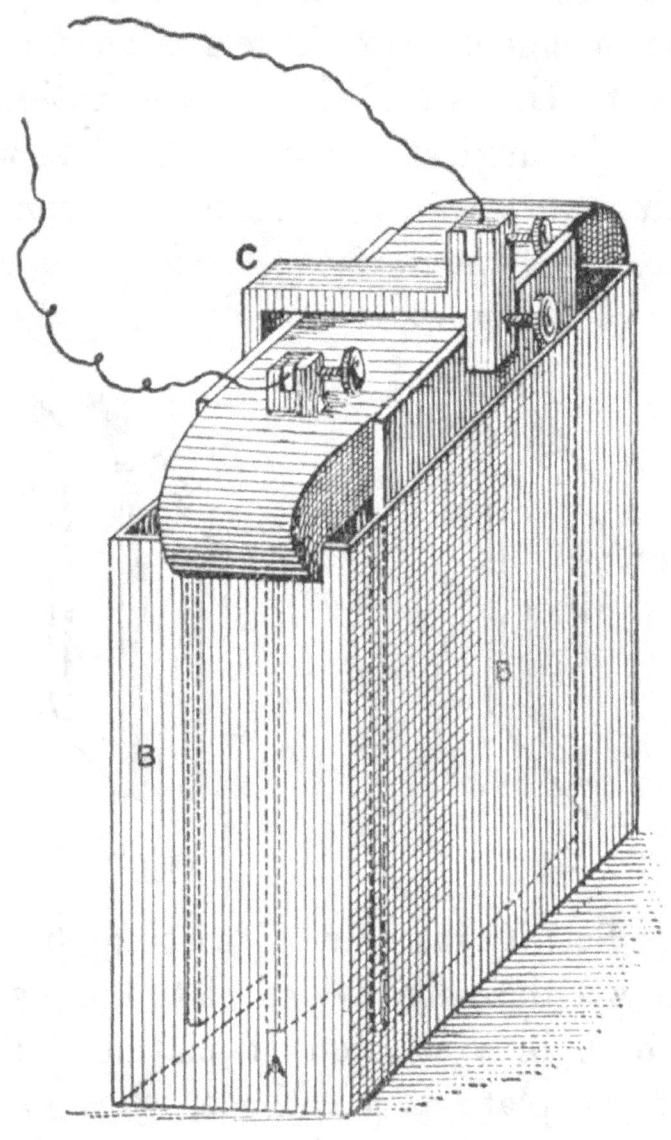

Fig. 37.

two zinc plates, electrically connected together by a metal clamp.   Fig. 35 shows the platinized silver sheet mounted in its frame, along with the terminal making connection with it.   Fig. 36 shows a similar frame sandwiched between the two zincs, which are held in position by the clamp *A*, which serves at once to connect the two zincs together, and as a terminal.   In Fig. 37 we have a complete cell.   For a very long time, practically until the advent of the dynamo, the Smee cell was the only one used for electro deposition, owing to the steadiness of its action.

§ 48. The next important improvement was that sug-

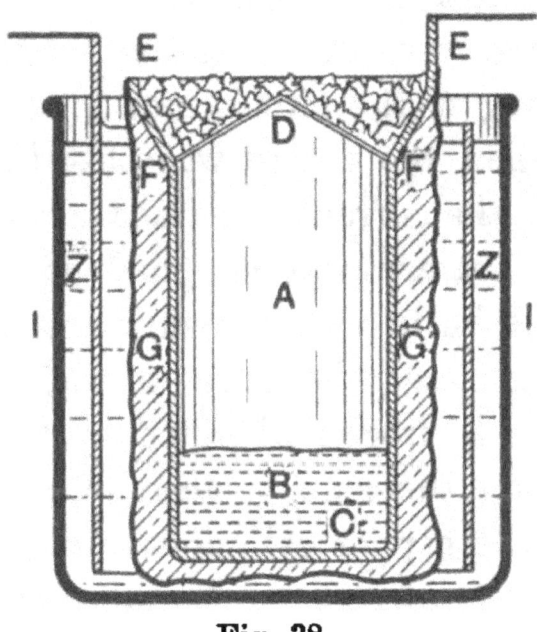

Fig. 38.

gested by *Becquerel* (1829).   It consisted of a hollow cylinder as shown in our Fig. 38, *A*, made of thin copper sheeting loaded with some sand *B*, and closed on all sides. The bottom *C* was flat, the top *D* conical, having over it a rim *E* perforated with numerous holes.   The whole cylinder was surrounded by a bladder *G* fastened to the rim *E* above the holes.   Upon the cone *D* were placed a few crystals of sulphate of copper, and a saturated solution of sulphate of copper was also poured into the top of the cone, and this running through the holes *E*, filled the space between the bladder and the cylinder *A*. Outside the bladder was a hollow cylinder of zinc. The whole was stood in a glass or porcelain vessel, which contained either dilute sulphuric acid or a solution of nitrate of zinc, or even of common salt. Two stout copper wires *P* and *W*, one of which was soldered to the zinc and the other to the copper, formed the two poles of the element. The advantage of this mode of construction was very great. In the first place, if the zinc were amalgamated little or no action took place as long as the circuit was interrupted. In the second place, when the circuit was closed and the battery was doing work, the current remained absolutely constant as long as the battery was supplied with its elements, namely, zinc, sulphuric acid, and sulphate of copper. The reason for this constancy is not far to seek : it depends on the fact that the hydrogen liberated by the action of the sulphuric acid on the zinc, instead of going to cling on the surface of the copper cylinder (thus polarizing it), on finding itself in the presence of the sulphate of copper, displaces the copper from its sulphate, giving rise to free sulphuric acid and to free copper, this free copper actually going to and being deposited on the surface of the copper cylinder exactly as the hydrogen would have done had not the copper sulphate solution intervened. But whereas the deposition of hydrogen would have quickly lowered the action of the cell by polarization, the presence of the copper exerts no such injurious effect. Besides this, the sulphuric acid liberated from the copper sulphate exactly balances and makes up for the amount taken from the original dilute sulphuric acid solution by the action of the zinc. As this cell is highly interesting both from a theoretical and a practical point of view, we make no apology for presenting once again to the reader's notice the chemical formula representing the changes which take place in the Becquerel cell :—

1st. In the zinc compartment—

$$Zn + H_2SO_4 = ZnSO_4 + H_2.$$

2nd. In the copper compartment—

$$H_2 + CuSO_4 = H_2SO_4 + Cu.$$

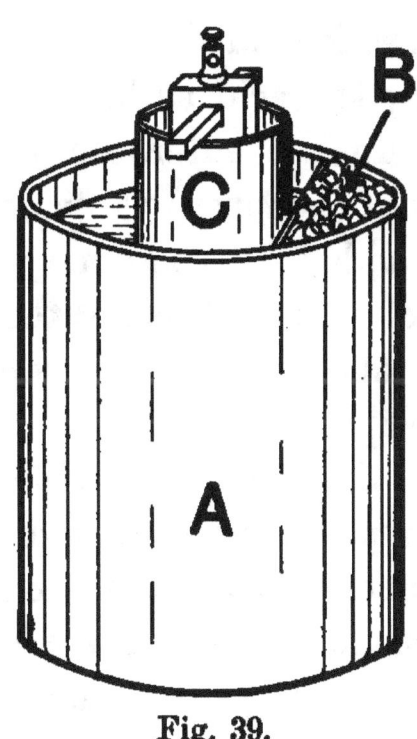

Fig. 39.

§ 49. Modifications innumerable followed upon the publication of Becquerel's invention; but the individual to whom the credit of having given the first practical working form to a battery of this description is Professor Daniell (1836), whose name consequently has become associated with the

' constant " battery.　In the most convenient form of the *Daniell* cell we have a cylindrical vessel of rather thin sheet copper *A*, Fig. 39, round the inside of which is a perforated copper shelf *B*, on which are placed crystals of copper sulphate.　In the centre of the copper pot or cylinder stands a porous cell *C*, which should extend about $\frac{1}{7}$ above the top of the copper cylinder.　Finally, in the middle of this porous vessel stands a rod or stout plate of well amalgamated zinc.　To set this cell in action a mixture of 1 part oil of vitriol with 20 parts of water by measure is made, and this when cool is poured into the porous cell, in which stands the zinc.　(The zinc should not be attacked by the acid as long as the electrodes are not in contact.)　In the outer compartment is placed a saturated solution of copper, to which $\frac{1}{24}$ of ordinary sulphuric acid is added (this is done to increase the conductivity of the solution), while on the shelf *B* is placed a quantity of crystals of sulphate of copper.

§ 50. The principle on which the action of a constant battery depends having been thus elucidated, other forms employing other depolarizers were soon brought to light. Of these we shall for the moment notice only the *Grove* cell, which was made known in 1839.　This, which has perhaps at once the highest voltage with the lowest internal resistance of any known double fluid battery, consists, as shown in our Fig. 40, of a square well-glazed earthenware or glass vessel *A*, in which is placed a sheet of zinc bent in the shape of letter ∪, as shown at *B*.　In this stands a rectangular porous cell *C*, containing a sheet of platinum *D*. To charge the Grove cell the amalgamated zinc plate is placed in the containing vessel, which is about three parts filled with dilute sulphuric acid (1 part acid to 12 parts of water by weight). The platinum plate is placed in the porous vessel, which is then filled nearly to the brim with strong nitric acid, and inserted between the limbs of the zinc plate. Owing to the readi-

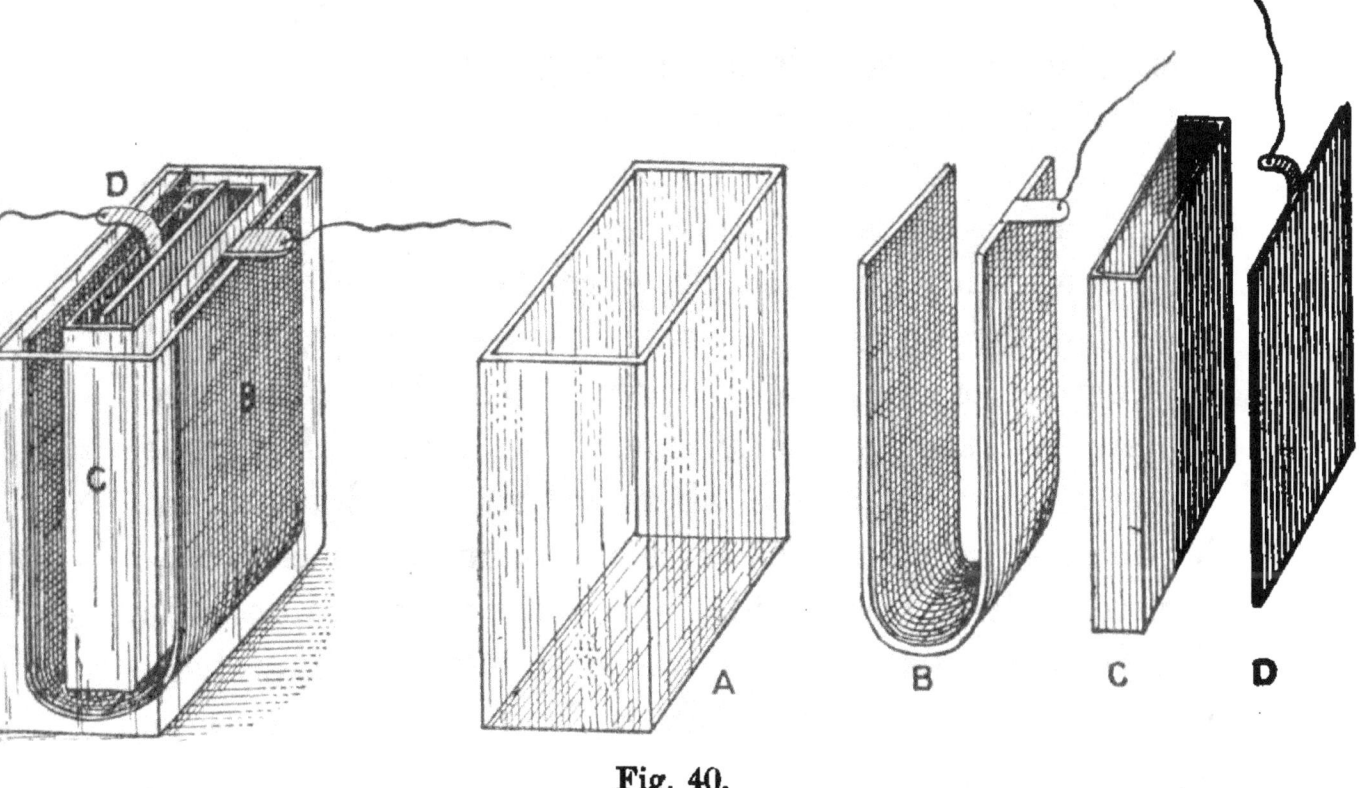

Fig. 40.

ness with which nitric acid parts with its oxygen, and the comparatively large quantity of this latter body which it contains, it acts as a most efficient depolarizer; consequently for heavy discharges continued through a considerable time, the Grove cell has few rivals among primary cells. The reaction which takes place in the Grove cell is shown in the annexed chemical equation :—

In the zinc compartment—

$$Zn + H_2SO_4 = ZnSO_4 + H_2.$$

In the nitric acid compartment—

$$H_2 + 2HNO_3 = 2NO_2 + 2H_2O.$$

Expressed in words this signifies that when one atom of zinc is acted on by a molecule of sulphuric acid, the zinc seizes on the sulphuric radical, with the production of zinc sulphate, liberating at the same time two atoms of hydrogen; and these latter, passing through the porous cell and finding themselves in contact with the nitric acid, seize upon a portion of its oxygen, forming therewith *water*, liberating at the same time nitrous acid. Although a most powerful battery, the Grove cell is open to the objection that the liberated nitrous fumes are not only most destructive of everything with which they come into contact, but also very injurious to the lungs. Its use is therefore only admissible where a thorough draught can be obtained, such as out of doors, or in a properly constructed "stink" cupboard. We now proceed to describe the various cells, which have from time to time been brought forward to public notice, laying particular stress on those which present interest, either from a theoretical or a practical point of view, beginning with those of the single fluid type, with no depolarizing substance added.

§ 51. *Pulvermacher's chain.*—Constituents, zinc and copper energized by vinegar held through capillary attraction. Arrangement: each cell consists of a hollow zinc cylinder about 1 inch long by ·16 inch in diameter; insulating rings are sprung on the ends of the zinc cylinder to insulate it from a surrounding copper cylinder, which is about ·3 inch in external diameter, and is perforated. The copper cylinder has extensions at its ends which are bent round so as to hook into and make contact with the zinc of the next cell. In this way a chain is built up with the cells in series. As a rule fifty couples are so joined and excited by dipping into vinegar, and then removing therefrom the vinegar is held by capillarity between the two cylinders. A single element is shown at Fig. 41 *A*, where *Z* represents the zinc and *A* the copper external sheath. Three such elements joined in series are shown at *B*. In another

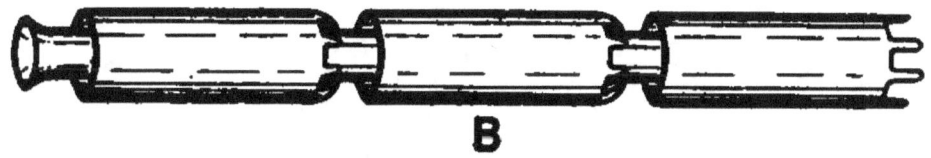

Fig. 41.

form a number of copper and zinc wires were wound on small pieces of wood, the connections being made as before between the terminal copper of one cell to the beginning zinc of the next. Vinegar was used because it was to be found in every household. Formerly this arrangement was largely used in medical treatment; now it does not meet with much favour.

Tyer modified the Smee cell by placing at the bottom of the cell a number of pieces of zinc, over which mercury is poured to effect electrical contact. The platinized silver plate or rolled cylinder has a leaden lug cast around its

H

upper extremity, which supports it just out of contact with the zinc pieces by resting on the rim of the containing vessel. The exciting fluid consists of 1 part by measure of sulphuric acid and 20 parts of water. To make connection with the contained zinc Tyer used strips of lead. These should be heavily varnished with shellac, except where they dip into the mercury and zinc.

§ 51*a*. The *Halse* cell is simply a *Smee*, in which one zinc only is used instead of two. Quite obsolete now (1901).

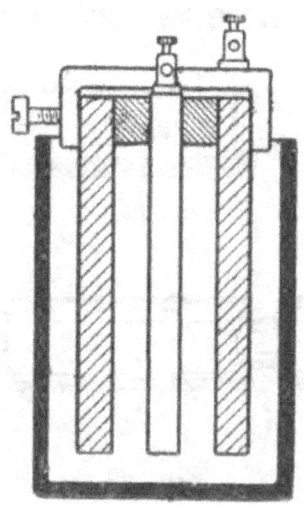

Fig. 42.

§ 52. *Baron Ebner* constructed a cell on the Smee plan, using platinized lead as the negative plate instead of platinized silver. This modification, while diminishing the prime cost of the battery, also diminished its E.M.F., owing to the fact of lead being nearer in the scale (see § 33) to zinc than silver.

§ 53. *Walker* (1859) modified the Smee cell by substituting platinized graphite plates for the platinized silver, and taking advantage of the, comparatively speaking, low price of graphite, used two graphite plates to one zinc, thereby gaining a much larger negative surface with a lower internal resistance, and consequently a larger current. This cell is shown at Fig. 42. Owing to the fact of the surface of the carbon or graphite plates being rough, and consequently presenting many points for the escape of the hydrogen, there was much less polarization in the Walker cell than in the original Smee; and the author, who had much experience in working with these cells while managing, in 1865, the Electric Bath Establishment belonging to Dr. Caplin, found them far superior in point of constancy to the ordinary Smee, the E.M.F. being, when dilute sulphuric acid was used as the excitant, ·66 volt as against ·47 volt of the Smee. A set of twelve elements connected in series (the number usually employed to furnish current for one bath) being sufficient generally to last for a week, the size of the carbon plates being 7″ × 5″, the zincs being a trifle smaller. The platinum is usually deposited on the carbons by immersing these plates in a weak solution of platinic chloride, suspending them for this purpose from a stout brass rod at the distance of about 1 inch apart, which rod is connected to the negative pole of a cell capable of giving about ·5 to ·8 volt, and using as the anode a large carbon plate also dipping into the solution, which is connected to the positive pole of the cell. It is practically of no use attempting to use platinum as an anode, as would theoretically be better than carbon, for the simple reason that the anode is so slowly dissolved that the strength of the solution is not maintained, hence it is better to work the solution to exhaustion, which can be recognized by the decoloration of the chloride solution. From careful tests made with roughened plates not platinized and plates platinized as above described, it would appear that the facility with which this cell depolarizes depends to a much larger extent on the roughness of the carbon surface than on the presence of the platinum black.

§ 54. Closely allied to the Walker cell above described is the *Maiche* cell (1879), of which we give an illustration

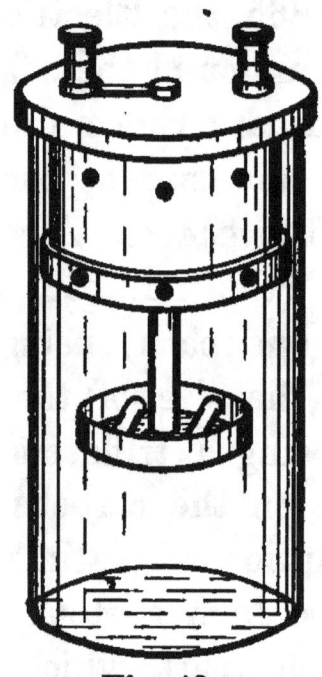

Fig. 43.

at Fig. 43. The constituents are zinc and platinized carbon, excited by water, acidulated with sulphuric acid, and saturated with either common salt or sal ammoniac. The arrangement as shown consists in fragments of zinc immersed in mercury standing in a tray, and fragments of platinized carbon contained in a perforated vessel either of carbon or of porous ware. If common salt be used the E.M.F. is about 1·25 volts; if sal ammoniac, the voltage is

higher, being 1·4 volts. Another arrangement is that of
a glass cylindrical vessel having an ebonite lid, to which is
fitted a porous pot perforated in the sides. This porous
cell is filled with fragments of platinized carbon, forming
the negative element. A platinum wire passing through
the ebonite lid serves to make a connection between the
fragments of carbon and a binding screw on the lid.
Through the centre of the lid passes an ebonite tube
reaching down to and supporting a porcelain basin containing mercury and fragments of zinc, which together
form the positive element. A second platinum wire passing through the ebonite tube connects the basin containing the mercury and fragments of zinc to the other binding
screws. The vessel was usually constructed to hold about
3 pints of liquid. In comparison with other cells this
is a cheap arrangement, as depolarization is effected by
the oxygen of the air. To allow the air to reach the
carbon freely, the porous vessel should dip into the liquid
to the depth of about 1″ only. The internal resistance of
a 3-pint cell is about 0·5 ohm. This cell, which is the
one figured, has been extensively used for telegraphic
work. *Helm* of Mulhausen (about 1851) brought out a
zinc-carbon cell in which a carbon rod stood upright in
the centre of a zinc cylinder. The excitant employed was
a solution of alum (hence this cell also acquired the name
of "Alum cell") to which common salt was added to
improve the conductivity. The Helm cell has had some
application for driving electric clocks; but its E.M.F. is
low and the internal resistance rather high. To drive 20
electric clocks, 16 cells in series were employed; and in

order to ensure uniformity in action, two cells, taken in
consecutive order, were refilled every week.

Edgar and *Milburn* (1891) used aluminium in a single-
fluid cell, and also in a double-fluid battery; of which a
notice will be found in its proper place. At the time of
its first publication, the price of aluminium was such as to
render its use in batteries prohibitive; now that it is prepared so cheaply, it might find some application. The
E.M.E., however, is not high. The following is the arrangement of the elements, as described by the patentees :—" A
simple fluid cell may consist of a carbon electrode, and an
aluminium electrode, which dip into a cell containing
ammonium chloride, hydrochloric acid and water; the
aluminium electrode being preferably placed in the centre,
while two carbon plates are placed to the right and left at
a suitable distance from the aluminium plate, these two
carbon plates being connected together externally, so as to
form one pole. The shape of the electrodes or poles, and
of the cells, may of course be varied, to suit various requirements." Two recipes are given for the exciting
solution, the first giving the more powerful effects.

1st.—Ammonium chloride	...	...	6 parts
Hydrochloric acid	...	...	3 „
Water ...	...	...	... 30 „

or—

2nd.—Ammonium chloride ...	...	4 parts
Hydrochloric acid	...	...	2 „
Water ...	...	...	... 30 „

§ 55. In *Switzerland* a very similar cell, the carbon not
being platinized, was for many years used for the Govern-

ment telegraphs. It consisted simply of an upright glass circular vessel, in which was placed a graphite cylinder about $5\frac{1}{2}''$ high and $3\frac{1}{4}''$ diameter. In the centre of the cylinder was suspended a rod or strip of well amalgamated zinc. The excitant was a saturated solution of common salt. The use of salt instead of sal ammoniac in these cells is to be deprecated, since, although salt is much cheaper, its E.M.F. is also much less, and the internal resistance is much higher when it is used. Many modifications of the last two cells have been suggested. *Sturgeon* (1840) used cast-iron in water, acidulated with $\frac{1}{8}$ of its volume of strong sulphuric acid (commercial oil of vitriol). *Münnich* (1849) used amalgamated iron; *Callan* (1855) employed a flattened cast-iron vessel as the positive in conjunction with a carbon negative, and hydrochloric acid as the excitant. Further on we will make mention of another iron cell due to Callan, in which iron is used as the positive metal in a two-fluid cell.

§ 56. To overcome the effects of polarization, due to the adherence of the liberated hydrogen to the surface of the negative plate, *Maistre* (about 1850), while using the ordinary zinc and copper elements in dilute sulphuric acid, kept one or both of the elements in motion, either, as he first suggested, by hand, or else, as finally adopted, by clockwork. By this means the current was kept fairly constant; and, according to Maistre himself, the E.M.F. is 1·08 volts, or a trifle higher than that of a Daniell. But the inconvenience of keeping up the motion by hand, and the expense attendant on using clockwork, render this form of battery unsuitable for practical work. Maistre's cell is interesting, however, from a scientific view, as it proves the pernicious effect of adherent hydrogen in lowering the output of the cell, and the benefit accruing from its immediate removal.

§ 57. Other means of getting rid of, or of facilitating the free extrication of the hydrogen gas from the negative plate without the use of special depolarizers, have been employed. Thus *Poggendorf* (1840) heated the copper plate in air until the colours which first appeared had passed away. This treatment oxidized the surface, and the oxide thus produced gave up its oxygen to the liberated hydrogen, converting it into water. With the same purpose in view, he immersed the copper plates first in nitric acid, and then washed them with water. He also found the covering the copper plates with a powdery deposit of copper by electrolysis facilitated the liberation of the hydrogen, by virtue of the large number of minute points created by this deposition.

In 1852 *Page* showed that if the copper plate be perforated with numerous burred holes previous to the deposition of the copper (which he preferred in the granular condition), the good results are largely enhanced. It is worthy of mention in this connection that the advantage of graphite ("retort scurf") as a negative element is largely due to the granularity or roughness of its surface. For this reason retort scurf is preferable to the smooth graphite blocks made artificially by moulding and baking carbonaceous mixtures in moulds, these latter giving off hydrogen much less freely. The employment of graphite (retort scurf) was first suggested by *Leuchtenberg* in 1845, and again by *Stöhrer* in 1849. The excitants used by these experimenters was a solution of alum in water.

We have already seen that Leuchtenberg (1845), Stöhrer (1849), and Walker (1859) employed graphite (gas carbon; retort scurf) as the negative element in their batteries. In 1894 Mr. H. T. Barnett brought out a battery in which the negative element consisted virtually of carbonized cotton velvet, or of carbonized fustian, backed with graphite. After carbonization, the velvet retains its " pile," and this, by virtue of the innumerable points it presents, so greatly facilitates the escape of the liberated hydrogen, that polarization does not take place to any injurious extent. Hence this battery is named the "Velvo-Carbon " or " Non-polarizing." As there are several points of interest connected with the construction, we present our readers with a brief description given by the inventor, and with illustrations of the various parts. The exciting fluid recommended for this form of cell (in which amalgamated zinc forms as usual the positive element) is a mixture of sulphuric acid with water, 1 to 6 being the strongest admissible, having a specific gravity of 1·180, while 1 to 12 (sp. g. 1·050) will be strong enough for general purposes. " Velvo-Carbon " is a generic name applied to battery electrodes, of whatever form or construction, whose surfaces are covered with carbon filaments. The advantages of these electrodes are that their surface is very great, and that the whole of that surface is in good electrical connection with the conducting support. They consist of velvet, attached to a conducting support of carbon or to a non-conducting support of porous earthenware. The back of the fabric itself may be prepared so that after carbonization it shall possess sufficient strength and electrical conductivity to allow of its acting as the sole conducting support to the filaments. The following summary indicates some of the uses to which the invention may be applied :—

Natural Depolarization.—The great surface and large number of points exposed render these electrodes more suitable for batteries of the Smee type (which employ only dilute sulphuric acid for the exciting solution) than the platinized silver usually employed. Their advantages are as follows :—(1) They are *cheaper*; (2) they are *not ruined by mercury* deposited from the solution; (3) they do *not* require *replatinizing*; (4) they do not polarize with hydrogen, even after very long periods of working; (5) when mercury polarized a short exposure to the air oxidizes the mercury and allows of its solution on the electrodes being again immersed.

Chemical Depolarization—Gassing.—With nitrate of soda or other gassing depolarizers, these "Velvo-Carbon" electrodes are also preferable to plain carbon; their many points liberate the gas freely, and their great surface is nearly all effective, as the evolution of gas causes a circulation of the depolarizer in the interstices of the pile. We do not recommend depolarizer solutions of any kind, as the simple sulphuric acid battery is cheaper to work, and in many other ways preferable.

Chemical Depolarization.—With non-gassing depolarizing solutions (for instance, the chromic) the advantages of "Velvo-Carbon" are only apparent so long as there exists undecomposed solution entangled in the pile. If great rushes of current are wanted at intervals, they may be employed. We do not recommend depolarizer solutions (more especially the chromic), as they are most costly and inconvenient.

Aërial Depolarization.—Carbonized corduroy immersed

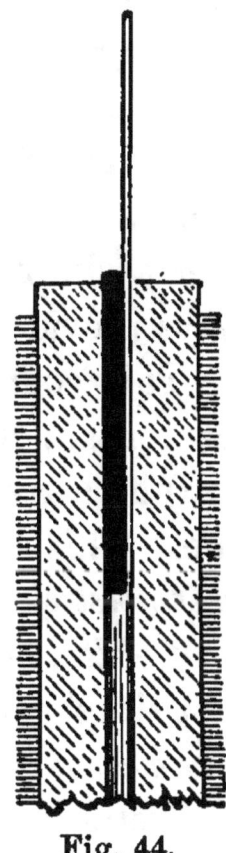

Fig. 44.

in air and situated on the outer surface of a porous jar forms a satisfactory negative electrode, the solution and positive electrode being contained in the interior of the porous jar. The pile of the "Velvo-Carbon" occludes oxygen freely from the air, and the porous nature of its backing allows the current to pass readily to and over the vast surface of the pile.

The facility with which the metallic connections become corroded, when used in conjunction with carbon plates or rods, has always been a weak point with cells employing graphite, etc.   Mr. H. T. Barnett remedied this by inserting a silver wire in a hole drilled in the upper edge of the carbon plate or rod.   Our Fig. 44 shows the upper extremity of a rod, fitted with its silver wire.   Speaking of these rods, the inventor says :—

"These are hollow carbon core rods of high quality with an adherent carbonized covering of short pile cotton velvet. They are suitable for use as the negative electrode of a battery, employing either a simple sulphuric or a nitric solution.   They are always provided with the patent pure silver wire connection.   This connecting wire is held in position (and in perfect contact with the wax-protected end of the hollow carbon) by a rivet-plug which expands after being driven home.   This connection is the *lightest, cheapest, cleanest, affords the best electrical connection, and is the least likely to get out of order*, of any yet devised."

The zincs should be not simply amalgamated by rubbing over the surface, but *immersed* in mercury during the operation : they then will retain their mercury to the last ; and the mercury can be recovered by distilling the battery mud and the spent zincs, at a gentle heat, and collecting the distillater.   The single cells of the "Velvo-Carbon" type are made up either with plates or rods. Two zinc plates with the "Velvo" between is very convenient for small work ; used with dilute sulphuric acid, a cell $3\frac{1}{4}''$ wide by $7\frac{5}{8}''$ long and $11''$ high will give 300 ampère hours, at a discharge rate of 7 ampères ; the E.M.F.

being 0·5 volt. Fig. 44*a* is from a photograph of four velvo cells arranged as a "plunge" battery. When using dilute acid the plates should be raised out of the cells as soon as current is not required. But solutions of caustic potash or caustic soda may be used instead of dilute sulphuric acid; in which case, as no action goes on when the circuit is broken, the elements may be left in the cell, even when

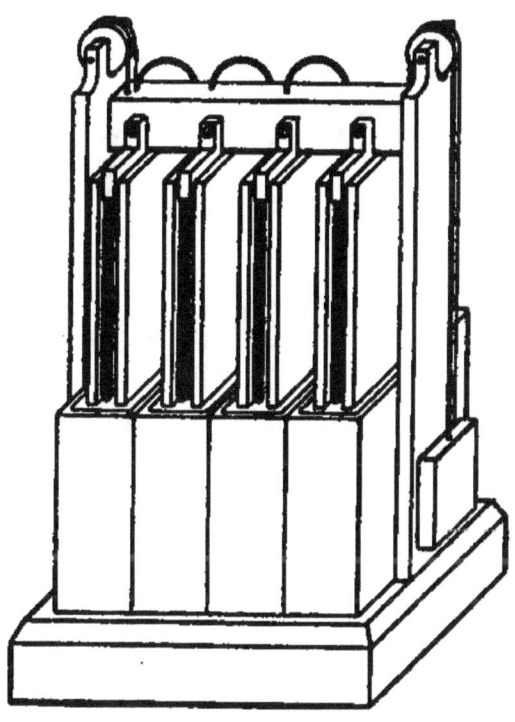

Fig. 44*a*.

the battery is not at work. Another form of this cell is the "Velvo-Carbon granule," illustrated at Fig. 44*b*.

This cell is specially designed for using up odd scraps of zinc. These may be used cut up small or granulated by being melted and poured into cold water from a height of about 6 inches from its surface.

To Charge the Battery.—Three-fourths fill the jar with cold dilute sulphuric acid, or other exciting solution, place some mercury in the porcelain cup, and then in it place some granulated zinc; replace the electrodes. The battery is then ready for use.

With dilute sulphuric acid as the exciting solution the E.M.F. on open circuit is about one volt, the output being about 35 ampère hours at a discharge rate of 2 ampères, the dimensions of the jar being 4⅛″ square, 7″ in height.

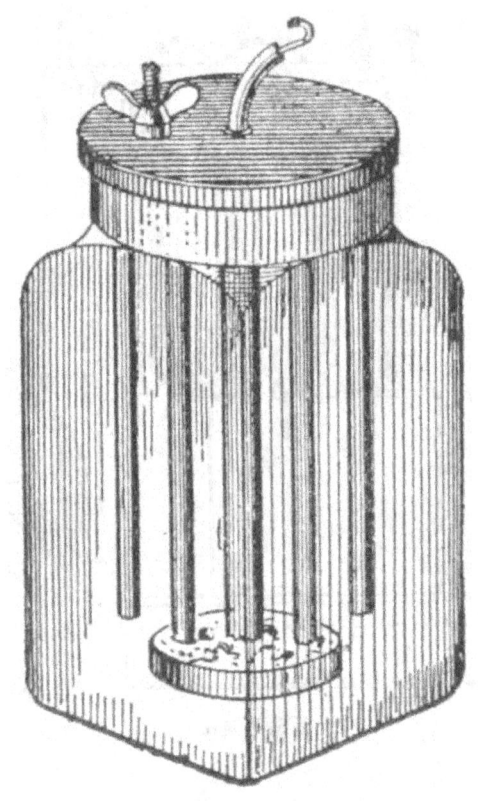

Fig. 44*b*.

This cell may also be used with a chromic or caustic alkali solution, when voltage would be 2 in the former case and 1 in the latter.

For heavy work, where higher voltage is necessary, the elements are conveniently mounted on a cross-bar, so as to form a large " plunge " battery. Such a battery is then suitable for the production of light, and for working electromotors, etc. They do not require charging frequently, and there is but little local action waste on the zinc plates. The solution is dilute sulphuric acid.

The consumption of zinc averages 1 lb. in each cell for every 320 ampère hours. With 1-a. lamps this is obviously

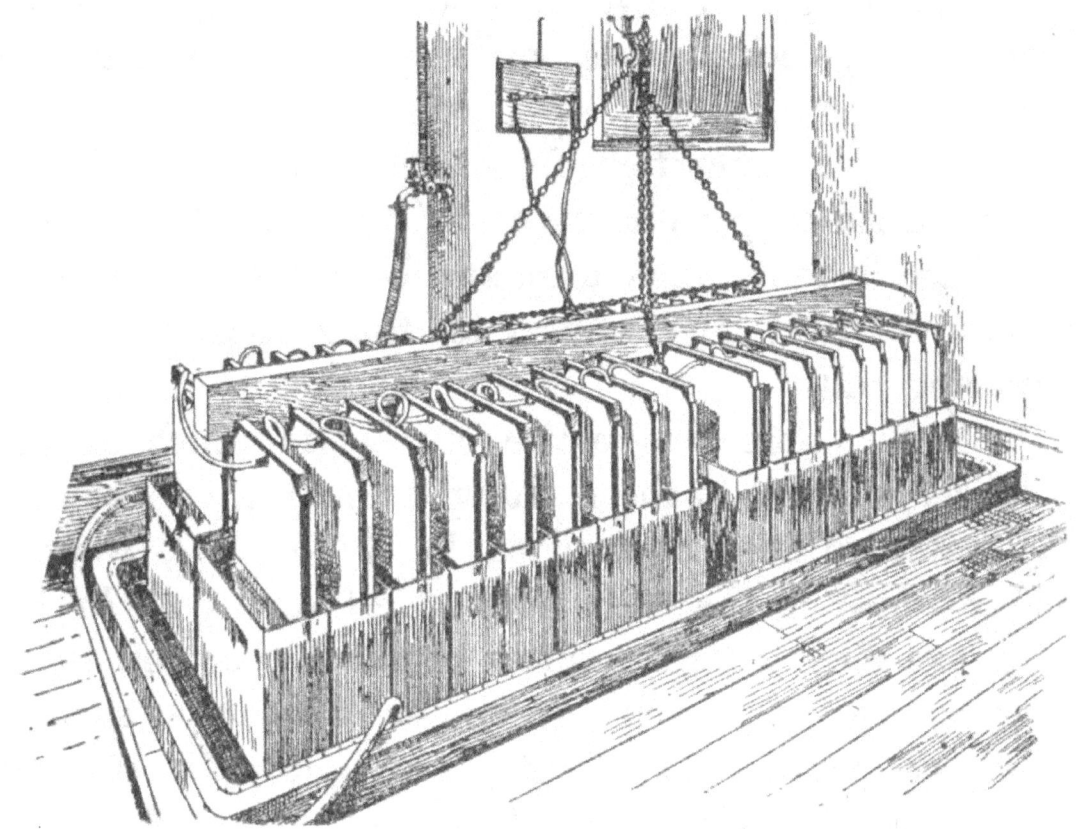

Fig. 44c.

320 lamp hours. The consumption of strong acid is $1\frac{3}{4}$ pounds for a like amount of current.

The cost of current will therefore be from about $\frac{1}{3}$ to $\frac{2}{3}$ of a penny per hour per 20 watt-lamp, varying with the price paid for the zinc plate.

At Fig. 44c we illustrate a 36-cell battery of this make suitable for general stationary work.

The 10-light battery is shown complete with a set of zincs in place and fitted up in a lead tray with water laid on to a tap at the back. The battery is raised and lowered as required by differential pulley tackle.

The battery is carried on a long bar which is provided with screw eyes to receive the suspending chain; across this bar, wedged in dovetail grooves, are the double-cell carrier bars, each bearing on either side a palisade of carbons.

The flexible connections from each carbon palisade are of fine strand copper wire covered with rubber tubing, and terminate in the suitable clamping bolts with attached fly nuts, which hold (and connect to) the zinc plates of the adjoining cell.

§ 58. We close this notice of cells in which no depolarizer is added to the exciting fluid by a reference to the water-batteries constructed by *De la Rive, Gassiot,* and *Rowland* in the early part of last century. In De la Rive's form a large number of small glass test-tubes were arranged in rows on a wooden stand, the upper shelf of which was perforated to admit the tubes (see Fig. 45), and copper and zinc wires joined together to form a ∏ dipped into water contained in the consecutive tubes. Gassiot modified this by mounting tiny copper or silver thimbles on the board, and from the edge of the first thimble carried a bent zinc wire to the middle of the next, the thimbles being filled with water as the excitant. Rowland modified and simplified this form of battery by arranging a number of very small zinc and copper plates, depending from a transverse supporting-bar, somewhat similar to the plunge form of trough-battery described and figured at § 39; with this difference, that the active pairs of zinc and copper were placed so close together, that if plunged into water, by capillarity some of the water would be retained between them on withdrawal from the fluid, while each pair of elements was removed from its neighbour to such a distance as just to prevent the adherence of water between them. Owing to the slow chemical action on the zinc, and to the enormous resistance of water, the current given by these water-batteries is extremely minute; on the other hand, each pair being so small, it is easy to

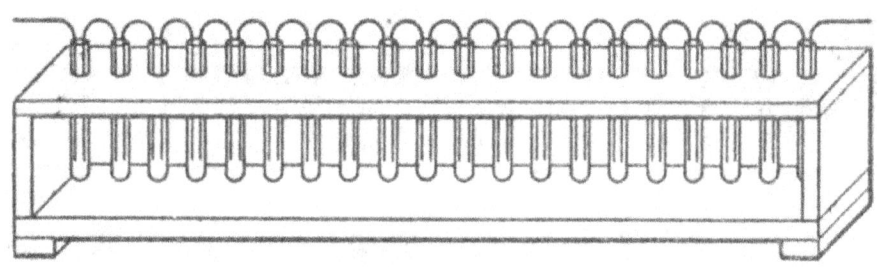

Fig. 45.

build up a set of 5000 or more in series in comparatively little space, and by this means obtain a very high voltage. Gassiot's great battery of 3520 couples constructed on this principle gave sparks of $\frac{1}{50}$ of an inch in length between its poles for several consecutive weeks. As before stated, Rowland's water-battery dispenses with tubes or thimbles altogether, and utilizes capillarity instead. The zinc and platinum or copper plates of each couple are placed very close together, while the couples themselves are more distant. On being dipped into and then removed from water each couple picks up and retains by capillarity a

I

little water between its plates, which water forms the exciting fluid. Many hundreds of couples can be mounted on a board, and the whole is easily charged by dipping into water and immediately removing. The battery then develops its full potential difference.

These batteries have been used for charging the needles of quadrant electrometers and for similar purposes. They polarize very quickly, and have a very high internal resistance; hence a large number of very small plates can be used without impairing their usefulness. The appearance of a Rowland's battery after removal from the

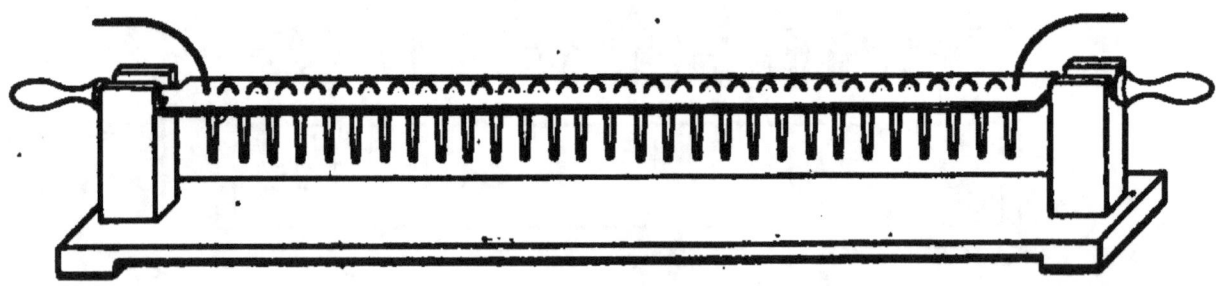

Fig. 45a.

water, and when supported on its stand, is given in Fig. 45a. A modification of *Rowland's* cell consists in employing little rods of *magnesium* instead of zinc. This is to be coated with paraffin-wax except at the lower extremity. As magnesium is more electro-positive than zinc, the E.M.F. is higher, and the current consequently larger. When studying the many modifications of the Daniell cell we shall see that Rowland also used magnesium as the positive element in a two-fluid battery.

§ 59. We now pass, by an easy transition, to those forms of the single-fluid cells in which there is present along with the excitant some substance which acts as a depolarizer. Although, as far as *constancy* is concerned (by constancy we understand the capacity of giving a fairly equal current as long as zinc, etc. and acid are in the cell), double-fluid cells, such as the Daniell, the Grove, etc., are far superior to any single-fluid cell, yet there are many cases in which a single-fluid cell, containing its own depolarizer, is to be preferred. One reason for this is, that the presence of the porous pot or other septum employed to keep the two fluids distinct from one another, gives rise to a very considerable *internal* resistance, so that the current is small; for instance, zinc and dilute sulphuric acid, pitted against a copper plate as negative, of 16 square inches in surface, at 1 inch apart, will give in a single cell about 1 ampère of current at 1 volt pressure; if we use the same elements in a Daniell cell the voltage remains the same, but, owing to the higher resistance of the porous cell, the current falls to about 0·2 ampère.

The attentive reader will have noticed at § 29 that we defined a depolarizer as a body that could absorb the hydrogen liberated from the acid by the action of the positive element, and thus prevent its adherence to the negative element. A knowledge of chemistry gives us the key to the choice of suitable substances. They may be bodies which themselves have a powerful affinity for hydrogen; such as chlorine, bromine, or iodine: or they may contain a large percentage of loosely combined oxygen, which oxygen, in the presence of the nascent [1]

---

[1] In chemical parlance, a body is said to be *nascent* when it is just released from its combination with another body.

hydrogen, unites with it, converting it into water: or, lastly, they may be metallic salts, consisting of an acid radical and a metal, of which the acid radical has a more powerful affinity for the hydrogen than it has for the metal. In this latter case, the hydrogen combines with the acid radical to form an acid, while the metal is set free.

§ 60. *Poggendorf* in 1839 suggested the employment of bichromate of potash as a most energetic depolarizer, when used in conjunction with the ordinary dilute sulphuric acid excitant. This salt consists of two atoms of potassium united with two atoms of chromium and seven atoms of oxygen, the latter being loosely held in combination. Expressed in chemical symbols, its formula is $K_2Cr_2O_7$. Poggendorf constructed his cell with a zinc and a carbon (graphite) element, standing in a glass jar, the composition of the fluid he used being:—

Bichromate of Potash   . .   100 grammes, or, 3 oz.
Boiling water . . . . . 1000 grammes, „ $1\frac{1}{2}$ pints.
Sulphuric acid (sp. g. 1·840)   50 grammes, „ $1\frac{1}{2}$ oz.

The sulphuric acid not to be added until the water has dissolved all the salt and has become quite cold. The effect of adding sulphuric acid to potassium bichromate in solution, is the production of free chromic acid and of potassium sulphate, as shown by the annexed equation:—

$$K_2CrO_7 + H_2SO_4 + H_2O = K_2SO_4 + 2H_2CrO_4$$

1 molecule of potassium bichromate	1 molecule and of sulphuric acid	1 molecule and cule of water	1 molecule give of potassium sulphate	2 molecules and of chromic acid.

As soon as the cell is put into action, and the fluid begins to act on the zinc, hydrogen is liberated; but instead of being set free in the form of bubbles, which in the ordinary course of things would go and adhere to the negative plate, it immediately seizes on the oxygen of the chromic acid, with which it unites to form water as shown below :—

$$3H_2 \quad + \quad 2H_2CrO_4 \quad = \quad 5H_2O \quad + \quad Cr_2O_8$$

| 3 molecules of hydrogen | and | 2 molecules of chromic acid | give | 5 molecules of water | and | 1 molecule of chromium sesquioxide. |

In the only original Poggendorf cell which the author

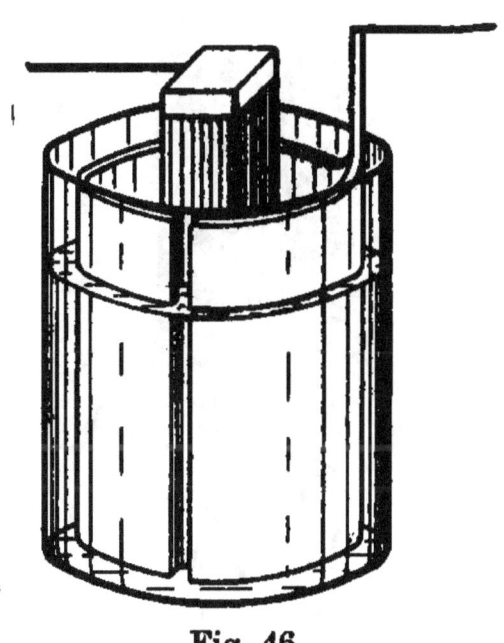

Fig. 46.

has seen, the zinc was in the form of a cylinder, the carbon being a block of graphite (gas carbon; retort scurf) standing upright in the centre, as shown in Fig 46. This form was probably adopted because of the difficulty of shaping gas carbon. The artificial production of graphite by mixing powdered gas carbon with molasses, moulding and then stoving the paste in moulds of suitable shape, was not then known.

§ 61. *Grenet* in 1841—very shortly after Poggendorf's suggestion—introduced the " bottle " form of bichromate cell (Fig. 47), which, from its portability, cleanness, and the ease with which the zinc can be removed from the active solution, without soiling one's hands or spilling the acid,

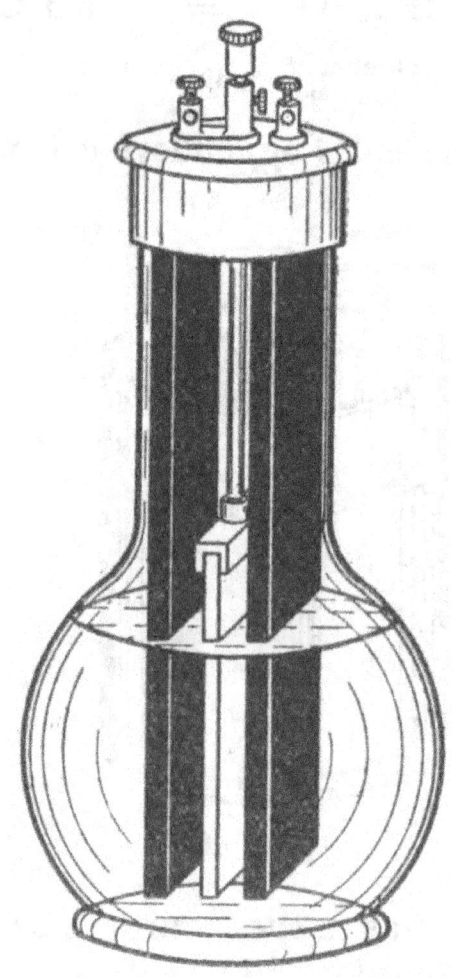

Fig. 47.

has long been a favourite with amateurs and others who require a powerful current for a short time, at intervals. As to the arrangement, the containing vessel is a glass flask or bulging bottle. The zinc plate is in the middle, and is capable of being raised out of the solution by means of a sliding brass rod, which can be clamped at any convenient point in its travel.

There is a carbon plate on each side of the zinc, and these two carbons are metallically joined at the top to constitute the positive pole. The solution acts on the zinc (which is well amalgamated) even when on open circuit, so that it is necessary to lift the zinc by sliding up the brass rod to which it is fixed when not in use. The solution is :—water 20 parts, potassium bichromate 3 parts, sulphuric acid 3 parts; all by weight. Annexed is the reaction :—

$$K_2Cr_2O_7 + 7H_2SO_4 + 3Zn = K_2SO_4 + 3ZnSO_4$$
$$+ Cr_2(SO_4)_3 + 7H_2O.$$

After some time two of the resulting salts combine together, thus :—

$$K_2SO_4 + Cr_23SO_4 + 24H_2O = K_2Cr_24SO_4 + 24H_2O ;$$

this latter salt, known as "chrome alum," separates out on the carbons in beautiful crystals of a deep amethystine red colour by transmitted light when left standing. This is detrimental, as the carbon plates become of much higher resistance in consequence. E.M.F. 2 volts, very nearly. Resistance of new cell, 1 pint size, with 4 square inches of negative surface, ·08 ohm. Such a cell will give a steady current of about 2 ampères for 3 hours, with one charge.

§ 62. Messrs. *Gaiffe, Trouvé* (1875), and *Tissandier* (1882) modified somewhat the form of the above, adapting it to being used as a battery of many elements coupled in series, and yet retaining the advantage of being able to remove the elements quickly from the exciting solution, or of immersing the same to a lesser or a greater depth, .

according to whether a small or a large current were required. To this end, the elements were arranged in pairs, suspended from a cross-bar of wood (previously rendered insulating and impervious to water by being immersed in boiled oil). This cross-bar, carrying the plates, was in its turn suspended by two stout gut bands from a windlass supported on two standards. This arrangement permitted the whole set of plates to be lowered into, or raised out of, a suitable number of jars

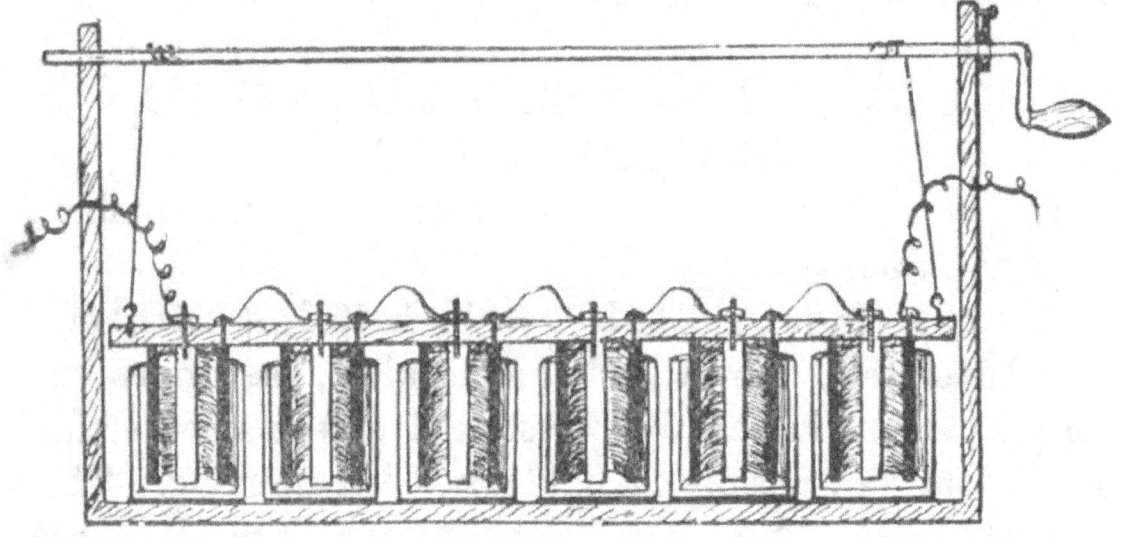

Fig. 48.

containing the exciting fluid. As the windlass was furnished with a ratchet and pawl arrangement, it was easy to stop the ascent or descent of the plates at any desired point. When the plates were withdrawn the acid dripped back again into the cells, without any fear of spilling or messing. We give at Fig. 48 an illustration of this particular form, now generally known as the "plunge" battery. For the sake of clearness, the figure shows the cells and the carbons in section only.

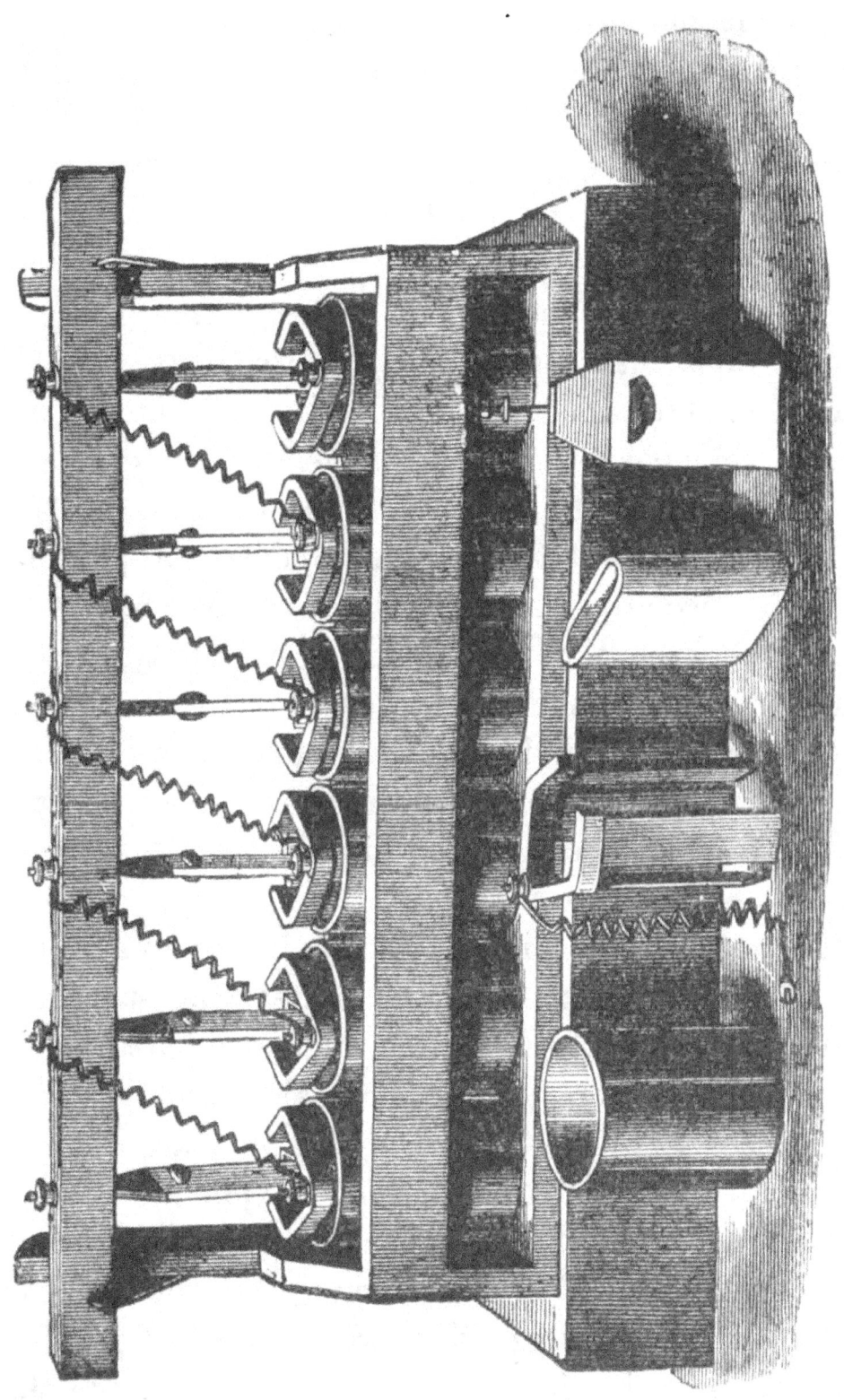

Fig. 48a.

"*Volta*" is the trade name given to a plunge-battery, very similar to above, in which four carbon plates, around which a lead collar is cast, surround the porous cell containing the zinc. See Fig. 48*a*. We are indebted to the International Electric Company for this illustration.

A modification of the plunge-battery, known as *Dr. Spamer's*, which has been pretty extensively used by medical men, and which is depicted in our Fig. 49, presents some points of interest. We have already mentioned, § 58, in speaking of the use of magnesium in a cell, that with a view to prevent the too rapid consumption of the metal, the device of coating the whole of the magnesium rod, with the exception of the lower extremity, with a resinous composition had been adopted. In the Spamer plunge-battery each individual cell (which may be of glass, porcelain, or ebonite) receives two rods, one of carbon and one of zinc. These are arranged in pairs, dependent from a transverse bar, as shown separately in the upper portion of our illustration; the number of pairs being equal to the number of cells to be used in the battery (30 such cells are shown in our cut). Each zinc is amalgamated before being mounted on the frame, and is also thickly coated all over, except at the lower end, with a mixture of resin and paraffin wax, melted together and applied while yet fluid. This precaution, by localizing the action of the acid, economizes the zinc. The bars carrying the elements can be lifted out of the cells, and fixed in any desired position by means of the rods and nuts working in the slots in the side of the box. A drawer contains all the cells, so that

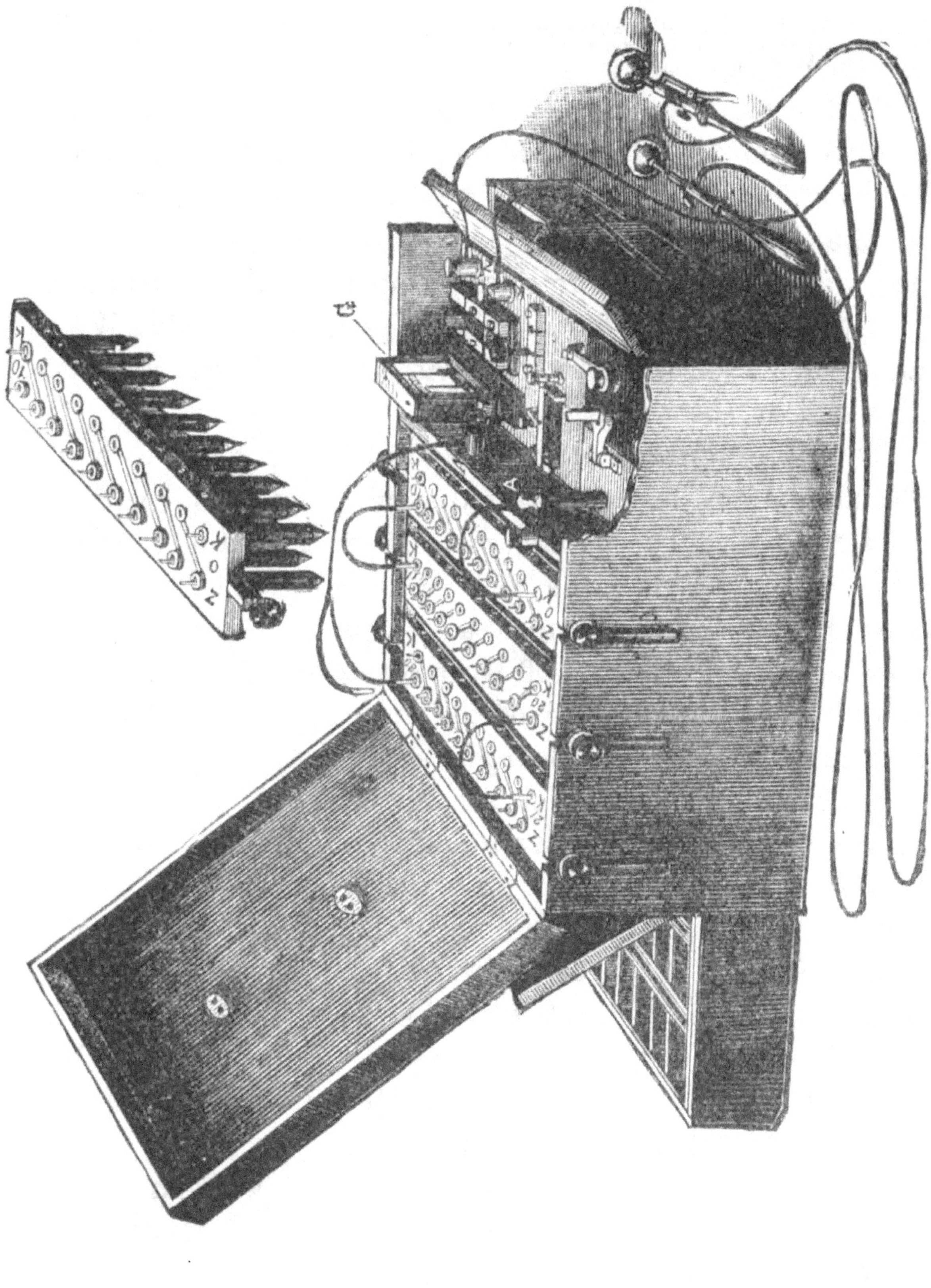

Fig. 49.

these can be slid out, for the purpose of charging, emptying, or replenishing.

Nearly every worker with the bichromate battery has adopted and given his name to his pet recipe for the making up of the exciting fluid; one adopting quantities of the bichromate and of the sulphuric acid in exact accordance with the chemical equivalents of these bodies; others allowing a preponderance of sulphuric acid. We annex a few of these recipes, with some of the results obtained by their employment :—

TROUVÉ'S.—Water 80 parts, powdered potassium bichromate 12 parts, concentrated sulphuric acid 36 parts; all by weight. The potassium bichromate is first mixed with the water, then the acid is added gradually, with constant stirring. As much as 25 parts of bichromate may be added to every 100 parts of water. The heating set up by the admixture of the acid with the water is sufficient to cause the bichromate to dissolve.

Used in conjunction with a zinc plate, with one carbon on each side, each exposing a surface of 15 square centimetres to the action of the fluid, the result was :—At the spurt 2 volts, internal resistance ·0016 ohm. After the spurt 1·9 volts, R. ·07 to ·08 ohm. On the short circuit d'Arsonval found this element to give 24 ampères for 20 minutes without polarizing; and a freshly-charged cell gave 180,000 coulombs (equal to 50 ampère hours) before the solution was exhausted.

TISSANDIER'S (1882).—Water 100 parts, potassium bichromate 16 parts, sulphuric acid, sp. g. 66°, Beaumé 37 parts; by weight, or what amounts to the same thing.

Water 1 pint, bichromate of potash 3 ozs., sulphuric acid 4½ ozs. The bichromate is to be reduced to fine powder, care being taken not to inhale the dust, which produces ulceration of the lining membranes of the nose. Part of the bichromate is dissolved in water at about 40° C. in an earthen vessel; the acid is then added, and the mixture is violently stirred until the whole of the bichromate is dissolved, the mixture is then allowed to cool down to 35° C. before using; below 15° C. the liquid works badly. A battery thus charged gives more than 1 horse-power, 2 or 3 hours, for a weight of about 200 kilogrammes, using 24 cells in series, and a Siemens dynamo as motor. One cell will give 100 ampères through an external resistance of ·01 ohm.

DELÀURIER'S formula. — Water 200 grammes, bichromate of potash 18·4 grammes, sulphuric acid 42·8 grammes. This is in accordance with the chemical equivalents of these bodies.

CHUTAUX'S formula (1868) is as follows.—Water 1500 grammes, bichromate of potash 100 grammes, bisulphate of mercury 100 grammes, sulphuric acid sp. g. 66, Beaumé 50 grammes.

ELECTROPOION FLUID is the somewhat fanciful name given to these bichromate solutions generally.

DRONIER recommended a mixture of 1 part of potassium bichromate with 2 parts of potassium bisulphate; the mixture keeps well in corked bottles, and a suitable solution can at any time be prepared by the addition of 3 parts of the mixed salts to 20 parts of boiling water. All these solutions must be allowed to cool before being used in the battery, otherwise the zinc is attacked too violently. They all are subject to the same defect, viz. that of giving rise to chrome alum during work, and this

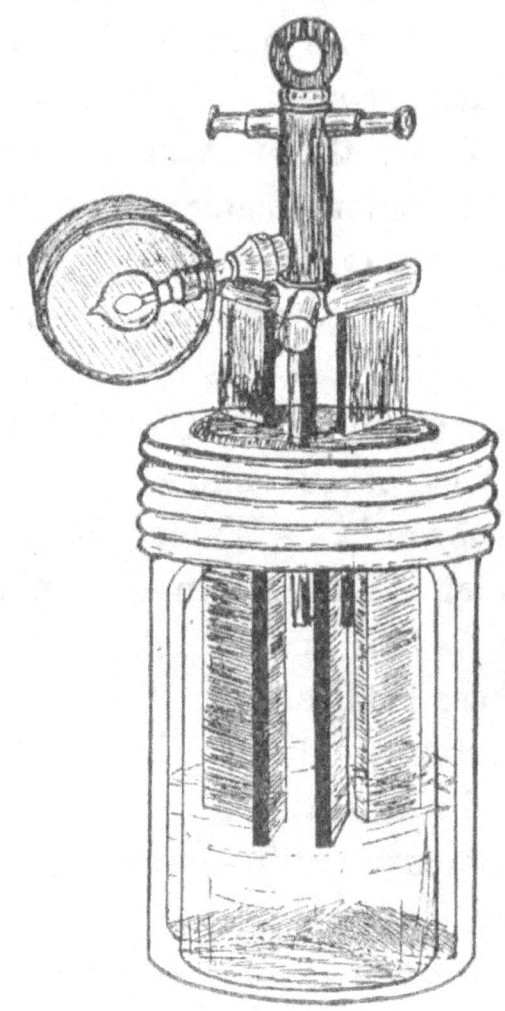

Fig. 50.

chrome alum, crystallizing on the carbons, greatly interferes with the efficiency of the cell.

Radiguet, in order to be able to light a 6-volt lamp with a battery not occupying much space, modified the bottle bichromate or chromic acid cell, by subdividing the containing vessel into three portions, with partitions of glass, ebonite or lead (a metal not acted on by the acid mixture). The zincs were made to slide up and down through slots in the cover, so as to be readily withdrawn from the excitant when the light was not required. The central lifting rod was constructed to carry the lamp and its reflector; the three pairs of plates were connected in series with one another. Beyond the neatness and compactness of this arrangement, there is nothing that can be called novel in it. Fig. 50 shows in outline the position of the three compartments and the general arrangement of the parts.

§ 63. The powerful effect which bichromate of potash has in seizing upon the liberated hydrogen is at once the strong point, and the weakness of all bichromate batteries; for although polarization does not take place until the chromium has been reduced to the state of sesquioxide (which can be recognized by the colour of the fluid changing from rich orange-red to a dull olive-green), yet the action rapidly falls off because, as there is no internal motion in the fluid (there being no bubbles of gas evolved), the solution in contact with the zinc soon becomes saturated with zinc sulphate, and consequently the chemical action ceases almost entirely. The slightest agitation, however, is sufficient to displace the zinc sulphate and allow the action to go on as before.

To overcome this defect, several devices have been proposed. *Dr. Byrne* of Brooklyn in 1878 devised a cell or battery of the general appearance shown in our Fig. 51, in which a current of air being blown through the fluid in the cells prevented the accumulation of the zinc sulphate in the proximity of the zincs. He patented a form of negative plate, which consisted of a sheet of copper covered on one side and on the edges with thin platinum foil (free from pinholes), and on the other with a sheet of lead, so soldered to the copper plate that this latter was entirely protected from the action of the acid while it retained its great conducting power. The back and edges

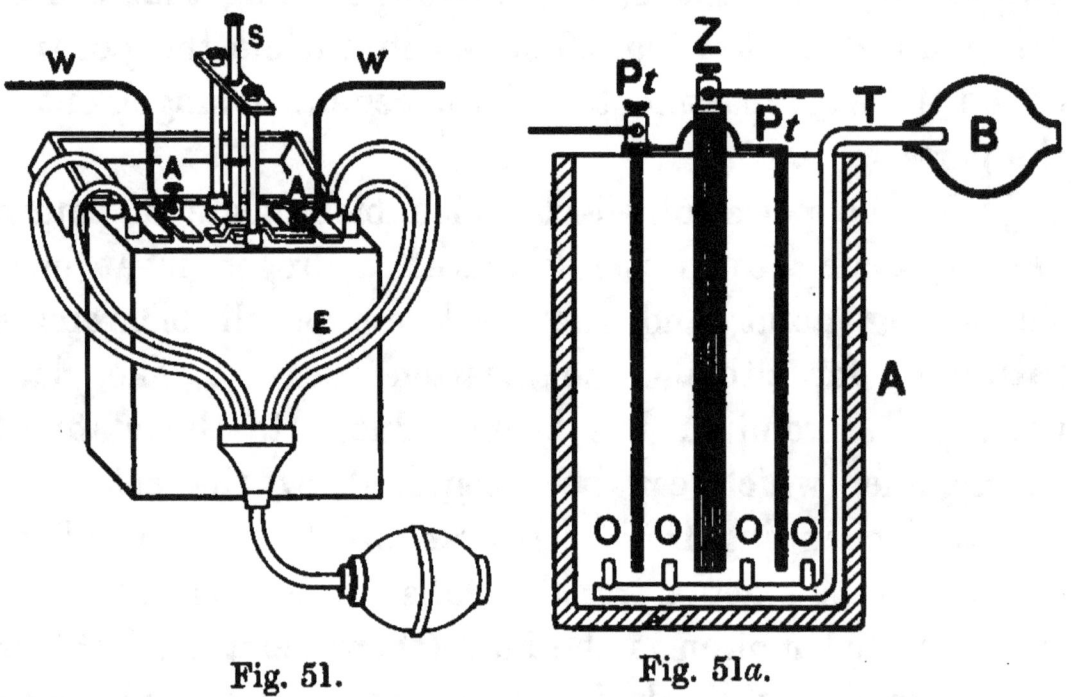

Fig. 51.   Fig. 51a.

are then painted over with asphaltum or any other acid resistant varnish. The platinum surface is then carefully platinized. Each cell contains two such plates, between which a zinc plate is suspended, so that when the combination is immersed into the exciting solution to such a depth that this latter reaches to within one inch of the top of the plate, a negative surface of 20 square inches is brought into action. Thus the copper core acts as a conductor only, the platinum being the real negative element, the lead serving as a protection from the acjd. The exciting solution recommended is made by dissolving 2 ounces of potassium bichromate in one pint of warm water. When the solution is cold, 4 ounces of oil of vitriol are added to it. (Calcium bichromate gives a much higher E.M.F. than the potassium salt.) The general appearance of the battery is seen in the annexed illustration (Fig. 51a). The cells are of ebonite or hard rubber, and the plates can be lifted out when not wanted. A thin lead or ebonite tube descends into the cell and extends horizontally along the bottom between the opposing plates, being perforated along the horizontal part, so as to allow a current of air which is pumped through it by a syringe, hand-pump, or rubber ball, to pass through the liquid and agitate it violently.

Ten cells gradually brought 32 inches of No. 14 platinum wire to a glowing red heat which ebbed and flowed with the cessation or renewal of the air supply. A brilliant arc was similarly maintained between two carbon points, and Spottiswoode's coil gave full intense sparks in air 18 inches long with air, and only 8 inches without air. The cell was much used for galvano-cautery.

R. Courtenay patented a somewhat similar form of battery, but obtained the motion of the fluid in the cells by arranging a series of gutta-percha pipes dipping nearly to the bottom of the cells, which pipes received fine streams of fresh bichromate solution from recipients placed above them, while near the top of the cells were arranged outlet pipes, which allowed the surplus acid to overflow into another recipient precisely similar to the

K

former. Since by once flowing through the cell the acid was by no means exhausted, a very steady current could be kept up by simply reversing the position of the upper and lower recipients as soon as the acid had flowed from the upper through the battery into the lower receiver. The author has had considerable experience with this

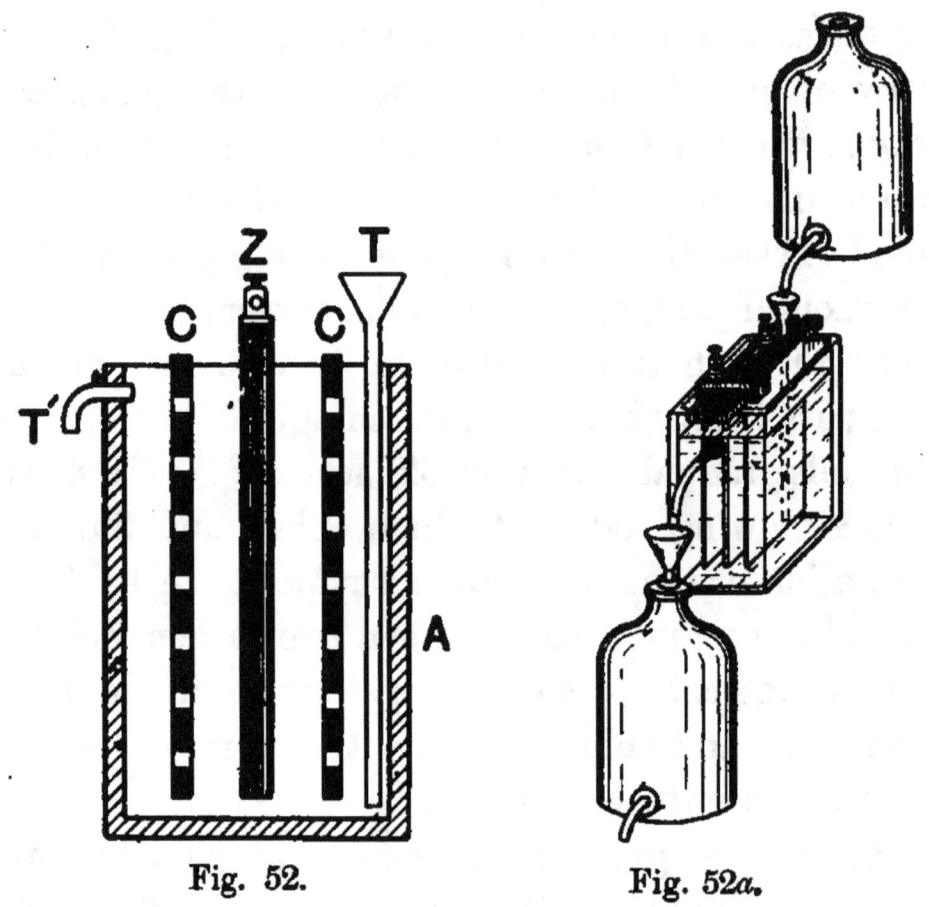

Fig. 52.   Fig. 52*a*.

form of bichromate cell, and found it very convenient where heavy currents are required for a long time, and where sufficient space can be allowed for the reception of the feeding and receiving vessels. With the substitution of chromic acid (which will be described later on) for bichromate of potash, so as to avoid the formation of chrome alum, this forms one of the most useful and powerful of the single-fluid type of cell. Fig. 52 gives a sectional view of such an arrangement.

§ 64. Mechanical means have also been applied to effect the same result. Fig. 53 shows a 4-celled bichromate battery in which the agitation of the fluid is effected by causing ebonite rings to agitate the fluid, by lifting and depressing a quadrangle which lies on the cover of the box that contains the 4 cells. This particular form of battery is noteworthy for the method by which the "plunge" effect is obtained. The battery, as shown, consists essentially of 4 glass cells, A A A A, about $2\frac{1}{2}''$ in diameter, standing on a tray T, from the centre of which rises a screwed and jointed rod R, by means of which it can be raised or lowered along with the 4 cells in the box B. The zincs and carbons are attached by means of long tanged binding screws to the lid L, and each element is connected in series to its neighbour by means of metal straps, the first and last, of course, forming the electrodes. The rod R, passing through the lid, being furnished with a nut which can run along the screw, enables the operator to *raise the cells* to the plates, either by drawing the rod up as far as the joint, or by screwing the nut down which raises the cells more gradually. In practice this will be found more advantageous than the older plan of lowering the plates into the cells. The device for setting up motion in the fluid consists simply in ebonite rings, E, E, E, E, which encircle the plates, and are attached to the ends of gutta-percha covered wires, W, W, the upper extremities of which pass through the lid of the box, and are soldered to the four corners of a flat square of wire *F*, that on being raised or depressed agitates the fluid in the cells, and thus prevents the accumulation of zinc sulphate round the plates. For the convenience of carriage, etc. the lid *L* is fastened to the box by means of eyes and catches,

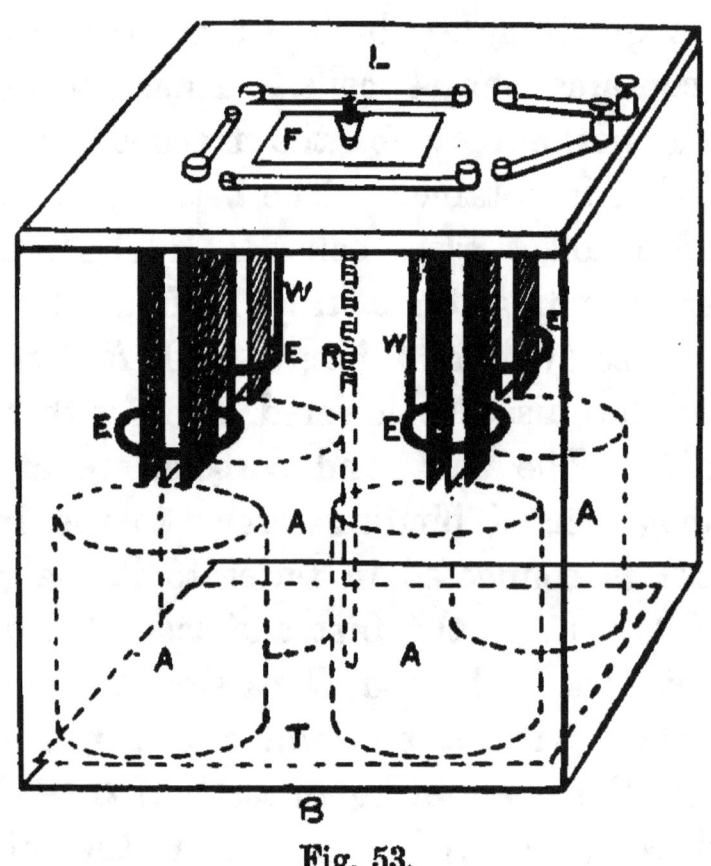

Fig. 53.

and is furnished with a central handle, not shown in the sketch.

§ 65. Another method for renewing the surface of the acid is by heating the recipient at the bottom, which causes a circulation of the fluid in the cell. This method was adopted by Sprague in 1874, and is thus described by that scientist in his work on electricity :—" Heat applied to the bottom of a cell has this effect, probably the heat adds to the E.M.F., but at all events it keeps it nearly constant by bringing fresh fluid to the plate. I have obtained a constant current from a bichromate cell, close up to exhaustion, by means of a small gas-jet under it. The heat also lowers the internal resistance, and so compensates for the lowering of E.M.F. due to the chemical change of the solution."

It is worthy of remark in this connection that, as long as there is a particle of bichromate of potash in the solution, the E.M.F. of a bichromate cell remains very fairly constant at practically 2 volts; but as the resistance of the fluid increases rapidly, as it becomes more and more charged with the products of decomposition, so the CURRENT falls off very quickly. It will be evident that the application of heat to the cells of a battery, though doubtless an effective means of keeping up the circulation of the fluid, is at once too inconvenient and too costly to admit of any extended practical use.

§ 66. Yet another means of keeping up this circulation has been proposed, and that is to utilize a portion of the current produced by the battery, to drive a small electromotor which keeps in movement one of the elements in the cell. It is, however, very doubtful whether anything is gained by this means, since the current wasted in the motor detracts from the useful current for outside work; the gain, if any, lies in the constancy of the effect.

§ 66a. *A. Wunderlich* and *O. Eigele* in 1888 modified the " Hare " cell by arranging the zincs and coppers in the form of discs, strung on a central spindle. All the coppers were connected together, as were likewise the zincs, but insulated from each other. By any suitable means, a rotary motion was imparted to the spindle, and in some cases provision was made to supply the acid from above and draw it off from below. By employing these two means conjointly, polarization was practically done away with. Yet more recently, it has been proposed to use the Hare's deflagrator in its original form, and simply to cause the central spindle to rotate, and to take the current off two metallic rings (one connected to the zinc, and the other to the copper element) by means of "brushes" as in a dynamo.

§ 67. As explained previously, the practical result of adding sulphuric acid to bichromate of potash is the liberation of free chromic acid. Hence it is evident that any chromic salt which will furnish chromic acid on admixture with sulphuric acid, or even chromic acid itself, can be used instead of the potash salt. As a matter of fact, any of the cells just described, belonging to the "bichromate" group, may be charged with either bichromate of soda or chromic acid. Both these are superior to the potash salt; the former, because it is far more soluble in water, hence the solution can be made much stronger (water will take up its own weight of the sodium salt), and also that the soda alum, *if formed,* does not crystallize out so easily. A convenient mixture, which can be kept in well-stoppered bottles, to be diluted when required for use with 4 or 5 times its weight of water, is known as "Voisin's Red Salt," and consists of 140 parts by weight of sodium sulphate, 686 parts of oil of vitriol, and 295 parts of potassium bichromate. The sodium sulphate is first dissolved in the heated acid; the powdered bichromate is then stirred in, finally the fused mass is poured into moulds, and when cold broken up and stored. It will be noticed that although a sodium salt is present in this mixture, it is not in the form of *bichromate*. Sodium bichromate can be had at about half the price of the potassium salt, and as it is more efficient it is much cheaper in use. A good working solution consists of—

Sodium bichromate . . . . 30 parts by weight.
Water . . . . 100 ,, ,, ,,
Sulphuric acid (sp. g. 1·845) . 23 ,, ,, ,,

Mix in the order given, allow to get quite cold before using the cell.

But far better than any of the above-named salts is commercial chromic acid,[1] which is now produced at a price only slightly above that of the potassium salt, and is far preferable; first, on account of its great solubility; secondly, because no sulphuric acid is wasted in separating the chromic acid from its base; and thirdly, because no chrome alum is formed during the action of the cell. For use in single-fluid cells, an excellent mixture is as follows :—

Chromic acid . . . . . 6 parts by weight.
Water . . . . 20 ,, ,, ,,
Sulphuric acid (sp. g. 1·845) 3 ,, ,, ,,
Chlorate of potash . . . $\frac{1}{3}$,, ,, ,,

[1] The pure "acid" is really a chromic trioxide CrO_3, but the commercial article contains 35 per cent. of sulphuric acid.

This addition of chlorate of potash is not obligatory, but it greatly conduces to the efficiency of the solution as an excitant and as a depolarizer: not only does the chlorate of potash part readily with its oxygen, but it causes the production of bubbling or "gassing," and this prevents the accumulation of zinc sulphate around the zinc element. As the result of some experiments made with a view to test the relative efficiency of these depolarizers, Sprague found that whereas a given cell working with the simple chromic acid solution gave only 6·4 ampère hours' work, a precisely similar cell, using the chlorate in admixture, gave 9 ampère hours.

§ 68. It has long been known that *nitric* acid, HNO_3, owing to the large amount of loosely held oxygen it contains, acts as a most energetic depolarizer; perhaps the most efficient yet known. Its employment was recommended by Gibson, Mason, Duffett, and Tweedale, but owing to its violent action on the zinc, its use has been chiefly confined to the "double-fluid" type of batteries, such as the Grove's and the Bunsen, which will be described in their proper place. Here we will only mention Schröder's single cell and battery, which are interesting, not only from the use of nitric acid, but from the very bold and original manner in which the inventor has ventured to place a number of elements in one cell,[1] and has succeeded in obtaining a terminal pressure, which, if not proportionate to the number of elements employed, is at least much greater than at first sight one would be led to expect.

[1] See § 47.

Mr. Hugo Schröder's single cell consists in a carbon-zinc pair, as shown in our Fig. 54. In order to obtain the very best results, he takes advantage of the fact that carbon becomes a better conductor when subjected to great pressure, and that this same pressure prevents to some extent the permeation of the carbon by the acid fluid. To obtain this pressure the usual terminals are

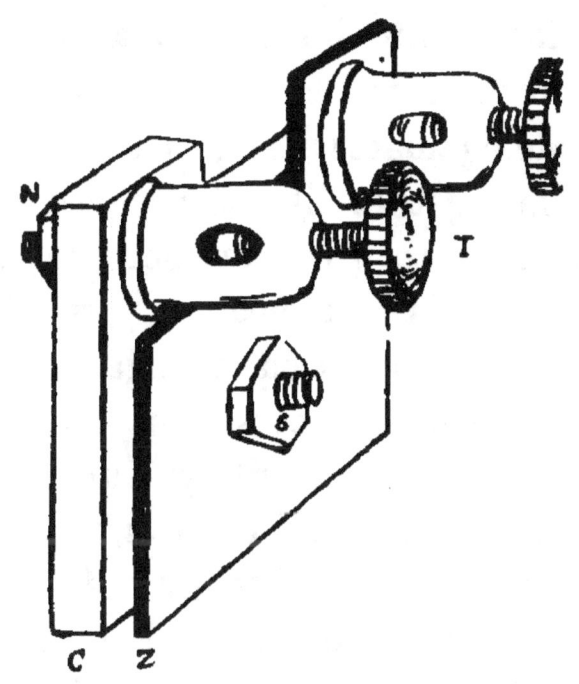

Fig. 54.

replaced by ones made of gun-metal very heavily gilt, and these are clamped to the zinc and carbon plates respectively by massive gilt nuts. By this means not only is much less current lost through imperfect contacts, which are the bane of batteries using carbon plates, but the terminals themselves remain always clean, the surfaces being practically incorrodible. Mr. Schröder found, as the result of many careful experiments, that it is a fallacy to suppose that absolutely pure zinc is the most suitable for battery purposes, and finds, on the contrary, that the presence of certain metals, notably aluminium, renders a plate immune to the action of acids when the circuit is open. Another great point in the duration of a zinc plate is uniform hard texture and freedom from crystallization, and this condition can only be secured by careful rolling or hammering the metal into very thin laminæ. For his zinc plates, therefore, Mr. Schröder procures a number of thin sheets, say of No. 24 B. W. gauge, amalgamates them heavily on both sides, then places them together like the leaves of a book, and finally subjects them to great pressure, say of 3 or 4 tons to the square inch. By this means a perfectly solid homogeneous zinc plate is produced, amalgamated throughout, and almost inert when immersed in the excitant, until the circuit is closed. The exciting fluid consists of sulphuric acid 6 parts, nitric acid 6 parts, water 20 parts by weight. Woodwork is therefore out of the question in mounting any such cell or battery. For this reason SLATE alone is used. A circular slate top with slots cut in it, to admit the passage of the carbon and zinc plugs, is fitted accurately, like a stopper, into the top of the glass cell forming the outer recipient, this top or cover being chamfered round the edge, so as to overlap somewhat the glass cell, the carbon and zinc plates are retained rigidly in position by means of a slate bolt and two slate nuts, a slate washer of the requisite thickness being inserted between the two plates, to keep them at the required distance apart. We do not show in the figure the glass cell or the slate cover; *Z* is the compounded zinc plate; *C* the carbon; *S* the slate nut and bolt; *T* and *T'''* the heavy gilt terminals; *N* the gilt nut compressing the carbon. An element exposing to the acid a zinc surface $2\frac{1}{2}''$ long × $3\frac{1}{4}''$ wide × $\frac{1}{8}$ thick, with a carbon plate of the same dimensions, but $\frac{3}{8}''$ thick, excited by the acid described above, gives a current through $6''$ of No. 16 German silver wire, of about 30 ampères, for an hour right off, which is sufficient to maintain that wire at a dull red heat for that period; and the current, far from decreasing, actually rises as the action goes on, until the nitric acid is consumed. Owing to the accuracy with which the cover fits the cell, very little, if any, nitrous vapours escape into the atmosphere, and Mr. Schröder has been successful in finding absorbents for these noxious vapours in the shape of certain metallic salts, such as potassic permanganate and ammonium molybdate, by the aid of which, if desired, these vapours can be entirely absorbed.

§ 69. Schröder's inventiveness did not, however, end here. By a careful adaptation of the internal resistance of his cell to the resistance of the outer circuit, in accordance with the requirements of Ohm's law, he has been able to raise the voltage or E.M.F. of a single cell almost at will, by increasing the number of alternations of elements immersed in the one recipient. This, perhaps, is the most extraordinary of the results obtained; and at first sight seems to clash with the received notions.

Schröder's "high tension cell," or battery, consists essentially, as shown in Fig. 55, of a number of pairs of zinc and carbon plates, prepared precisely as described in the foregoing section, fitted accurately, and subjected to pressure, so that each zinc and carbon pair make perfect electrical contact along the whole of their inner surfaces. This arrangement does away with one great source of trouble with ordinary cells, viz. there are no clamp connections to fit and get corroded through the acid being sucked up by the carbon. To avoid the creeping of acid from the last carbon plate to *its* terminal, this last carbon plate is likewise clamped to a longer zinc plate Z', to

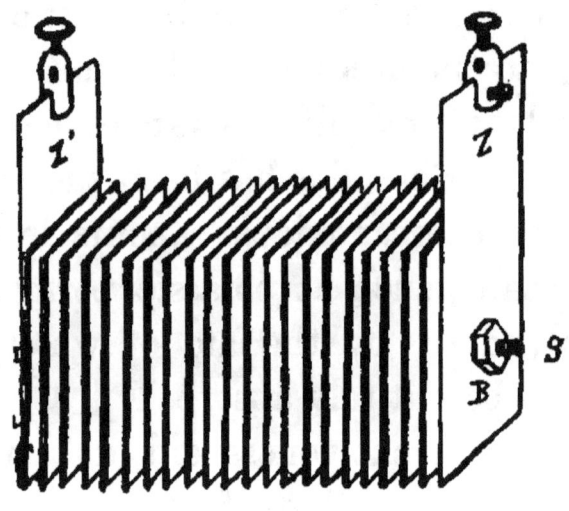

Fig. 55.

which the + terminal is attached. A long, massive slate bolt, tapped and fitted with slate nuts at both ends, passes right through the centre of *all* the pairs, each pair being separated from its neighbour by a thin slate washer. The intelligent reader will immediately perceive, that on immersing this compound arrangement into a vessel containing the exciting fluid (of the composition given above), galvanic action will be set up, and he will doubtless imagine that all, or nearly all the energy thus liberated will be consumed *in the cell itself* by the short circuit effect of the excitant.   Such is not, however, found to be the case in practice; and in fact, if the resistance of the outer circuit be kept low in proportion to that of the internal resistance in the cell, a very large current of high terminal potential is available for use in the outer circuit. Our figure represents the elements of one of these cells, with 20 alternations of zinc-carbon plates, size about $6'' \times 4''$, with the longer electrode plates, and heavy gilt terminals.   The author was present at the trial of such a cell, and found it capable of giving a current of 26 ampères, at about 36 volts pressure, which pressure falls in proportion to the load put on the outer circuit.   The internal resistance of this compound cell is about 1 ohm. The current is remarkably steady for about $1\frac{1}{2}$ hours—that is to say, until the zincs are consumed.   About 1404 watt hours can be got out of such a cell.   To obtain the very best results, it has been found advisable to feed the cell with a continuous fine stream of acid, which having circulated in the cell is allowed to overflow from the cell into a tank beneath.

With regard to efficiency, compactness, and ease of manipulation, this battery holds a very high position, and it is interesting from a scientific point of view; but the cost of preparing the zincs in the special manner described, the turning up, tapping the slate bolt and nuts, and the heavy gilt terminals, render it very expensive. For certain purposes wherein the use of a dynamo would be inadmissible, such as the production of a brilliant light (for optical lantern work at lectures, etc.), it has found pretty extensive application, when used in conjunction with Schröder's semi-incandescent lamp, of which we present an illustration at Fig. 56.   A detailed description of this lamp, which will give a light of 320 actual candle-

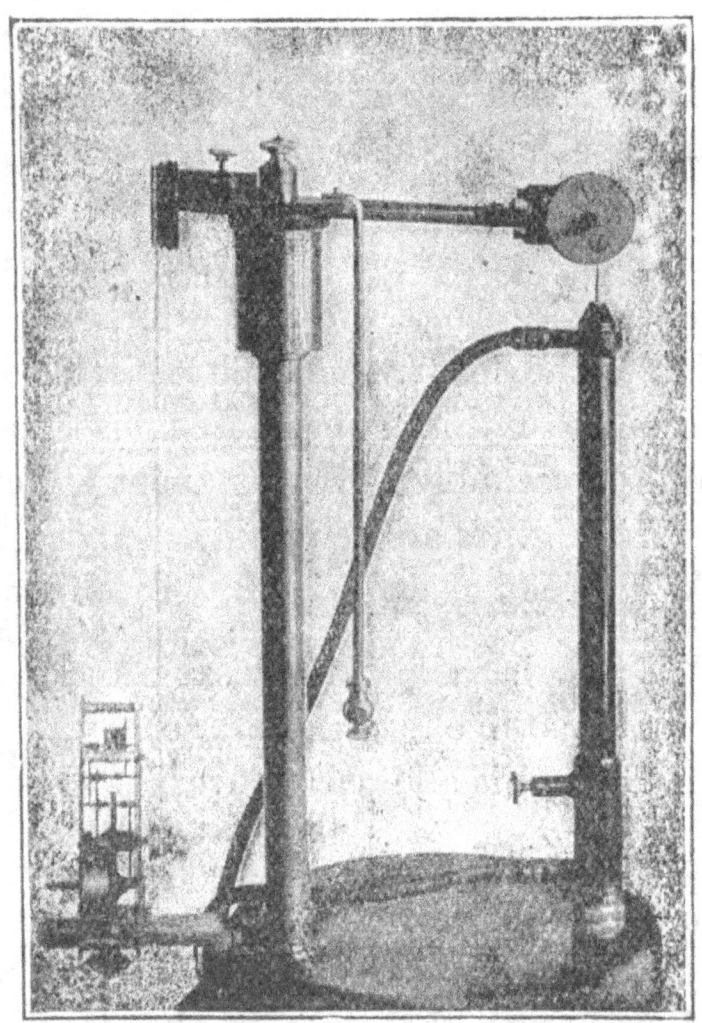

Fig. 56

power, with a current of 15 ampères at 6 volts only, will be found in the *English Mechanic*, vol. 62, p. 306, Nov. 22, 1895.

§ 70. Very similar in its effects as a depolarizer to

nitric acid, is nitrate of soda in admixture with sulphuric acid. This mixture has been employed either alone or in conjunction with bichromate of potash, of soda, of lime, or lastly with chromic acid. When sulphuric acid is added to potassium nitrate, or in fact to any of the nitrates which have been suggested for use in batteries, the sulphuric acid seizes upon the base, to form a sulphate, while nitric acid is set free. This reaction is shown in the following equation :—

$$\underset{\substack{\text{Potassium} \\ \text{nitrate}}}{KNO_3} \quad \underset{\text{and}}{+} \quad \underset{\substack{\text{Sulphuric} \\ \text{acid}}}{H_2SO_4} \quad \underset{\text{give}}{=} \quad \underset{\substack{\text{Hydrogen} \\ \text{potassium sulphate}}}{HKSO_4} \quad \underset{\text{and}}{+} \quad \underset{\text{Free nitric acid.}}{HNO_3}$$

Hence the final result, as far as regards the action in the cell or battery, is precisely similar to that of adding an equivalent amount of nitric acid as the depolarizer. All the objections that militate against the employment of nitric acid, apply with equal force to the mixture mentioned above. As soon as the nitric acid begins to be decomposed by the nascent hydrogen, so soon are the irritant and noxious fumes of nitrogen trioxide given off. This being the case, it is evident that no real advantage is gained, in point of efficiency, or even of freedom from the deleterious action of the liberated fumes, by the employment of a nitrate instead of nitric acid. Even as regards precise cost, there is very little to choose. Nitric acid of sufficient purity for battery purposes can be obtained at about half the price of potassium nitrate. If sodium nitrate (*cubic nitre*) be used, this, being much cheaper, brings the cost down to about the same as commercial nitric acid. Many experimenters have given recipes for depolarizers containing nitrates : the only one we shall mention here is the "*Fermoy*" cell, described by its inventor in 1890. The cell is of the ordinary zinc-carbon type, and may be advantageously constructed with the zinc sandwiched between the two carbons. These latter, after connection to the terminals, should be heavily coated with hot melted paraffin wax at their upper extremities (as should also be the lower portions of the terminals). The exciting fluid consists of a mixture of nitrate of soda, bichromate of potash-water and hydrochloric acid, in the following proportions :—

Nitrate of soda	...	...	...	...	3 oz.
Bichromate of potash...		...	...	...	3 „
Water ...	...	...	...	...	20 „
Hydrochloric acid	...	...	...	...	3 „

As long as there remains any free chromic acid in the solution, nitrous fumes are not given off; but as soon as this has taken place, and the liberated hydrogen begins to attack and decompose the nitric acid of the nitrate of soda, the usual fumes appear. The production of these fumes is accompanied by a slight effervescence, or "gassing," and this is by no means a disadvantage from an electrical point of view, as the motion thereby set up in the mass of the fluid prevents the accumulation of the zinc salt round the zinc plate. The nitrates are therefore often called " gassing depolarizers." The author has experimented with this cell, and found that the E.M.F. is pretty constant at 1·9 volt for about four hours consecutively, and the current also is fairly constant. He modified somewhat the formula for the depolarizer, and obtained even better results with a solution compounded as below :—

Nitrate of soda ... ... 10 parts
Chromic acid [1] ... ... 10 „
Hydrochloric acid ... ... 3 „
Water ... ... 20 „

With this solution the initial E.M.F. was 2·2 volts, and the current on the short circuit, using plates exposing 16 square inches of surface to the acid, was nearly 25 ampères. Two of these cells coupled in series lighted a 4ᵥ 0·25ₐ lamp brilliantly for six hours, after which the filament became appreciably fainter in its glow, and in nine hours was only dull red.

§ 71. Permanganate of potash, and the corresponding sodium salt, have both been tried as depolarizers. The formula of the former is $K_2 Mn_2 O_7$, and of the latter $Na_2 Mn_2 O_7$, so that owing to the large amount of oxygen they contain, they should be equal, if not superior, to the bichromates in action. Unfortunately their efficiency is marred by the rapidity with which they are decomposed on the addition of any acid; and also to the fact, that while giving up their oxygen readily, they (or rather the permanganic acid present) are reduced to the condition of manganese dioxide, which adheres as a brown sludge to the surface of the carbon plate, filling up its pores, and rendering it practically non-conducting. Until this takes place, however, the permanganate cell is most energetic, its E.M.F. being very nearly three volts. The first effect

---

[1] Ordinary commercial, which contains about 30% of sulphuric acid.

L

of adding sulphuric acid to a solution of a permanganate is to liberate permanganic acid to form a sulphate, thus—

$$Na_2\ Mn_2\ O_7\ +\ H_2\ SO_4\ =\ N_2\ SO_4\ +\ H_2\ Mn_2\ O_7$$
Sodium Permanganate and Sulphuric acid give Sodium Sulphate and Permanganic acid.

But this last body is not stable; it almost immediately resolves itself into free oxygen, water, and manganese dioxide, thus :—

$$H_2\ Mn_2\ O_7\ =\ H_2O\ +\ 2\ Mn\ O_2\ +\ O_2$$
Hydrogen Permanganate gives　Water　and Manganese dioxide and Oxygen.

As an experiment, to demonstrate the great activity of the oxygen thus presented in the nascent state, the following solution has been recommended for use in a zinc-carbon cell :—

Sodium Permanganate...	2 oz.
Water	1 pint
Sulphuric acid	3 oz.

All by weight. First add the sulphuric acid to the water : allow to cool. When *quite cold*, add the sodium permanganate, and stir with a glass spatula until dissolved. To be used at once. As before stated, at the spurt, the E.M.F. of such a cell is very high, reaching three volts ; and the conversion of the nascent hydrogen into water is so perfectly effected, that there is absolutely no extrication of hydrogen bubbles, even if the proportion of sulphuric acid be largely increased over that given above ; but owing to the rapid deposition of the reduced manganic oxide on the surface of the carbon, the internal resistance quickly rises, so that the current soon falls.

§ 72. *Marié Davy* was, we believe, the first to employ sulphate of mercury, $Hg_2\ SO_4$, as a depolarizer. Originally the inventor used this mercury salt in a "double-fluid" cell; Fitzgerald and Schanschieff modified the arrangement so as to render it serviceable in the single-fluid type. In the single-fluid Marié Davy cell, illustrated at Fig. 57, we have a square ebonite cell or trough *A*, into the bottom of which is inserted a square slab of graphite *G*. From this rises, embedded in the ebonite, a wire *W*, which forms one of the electrodes. The cell has, on two of its sides, ebonite ledges *L L*, on which can rest a stout zinc plate *Z*. Either connected to the zinc itself, or else lying

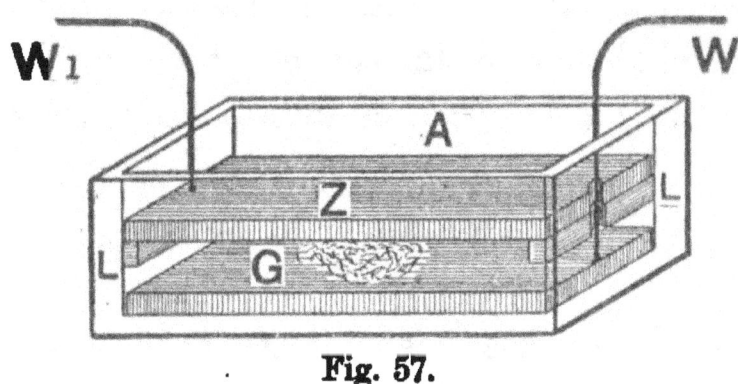

Fig. 57.

along the top edge of one of the ledges, is a second wire *W'*, which forms the other electrode. To set the cell in action, a small quantity of the mercurous sulphate is sprinkled over the carbon *G;* water is then poured in until it reaches the top of the ledges *L*. The zinc plate is now placed resting on the ledges, and in contact with the water. A portion of the mercurous sulphate is dissolved; the zinc seizes upon its sulphuric acid, setting up galvanic action thereby, at the same time forming zinc

---

[1] In the figure, the outer ebonite case is represented as if transparent, so as to show the relative position of the zinc, the carbon, and the mercurous sulphate.

sulphate, and setting free metallic mercury. A portion of this mercury goes to amalgamate the zinc, another portion falls on the graphite plate, where it does no harm, as it is a good conductor, electro-negative to zinc. The E.M.F. of such a cell is about 1·5 volts. The cell is fairly constant, but the internal resistance is pretty high. The reaction that takes places in the mercuric sulphate cell, as far as the chemical changes are concerned, are as shown below :—

$$Zn + Hg_2 SO_4 = Zn SO_4 + Hg_2$$

Although the above represents the final result of the action, it is very probable (as water is present during the reaction) that the operation takes place in two steps, thus :—

$Zn + H_2O = Zn O + H_2$; then $H_2 + Hg_2 SO_4 = H_2 SO_4 + Hg_2$; and finally $Zn O + H_2 SO_4 = Zn SO_4 + H_2O$.

This form of cell has been pretty extensively used for exciting small pocket coils; as the fairly high E.M.F., the absence of any corrosive fluid, and the freedom from fumes, also from local action when the circuit is broken, render it very convenient for this purpose. In this case the cells are generally put up in pairs, as shown in section at Fig. 57a, in order to get a higher E.M.F. The trough is made of ebonite, and is divided in the centre by an ebonite partition, thus forming two distinct compartments or cells. The carbon C in the left-hand cell is connected to a platinum wire, which in its turn is attached to an external contact piece B, against which, when the little battery is in its place in the coil-case, presses against the spring R, that is in connection with one terminal of the coil. The zinc Z rests on one side on the ebonite ledge t, and on the other on a platinum wire P, which wire passes through the central partition, and makes good metallic contact with the carbon C' in the right-hand cell. In this right-hand cell the zinc rests at t' on an ebonite ledge, and at P' on a platinum wire, which passes through the cell to the outside at A, where, when in position, it is pressed by the spring R' which constitutes the other electrode of the coil. To charge this battery, the zincs $Z Z'$ are lifted off the ledges, a little basic sulphate of mercury placed at the bottom of each cell. Water is then poured in, just above

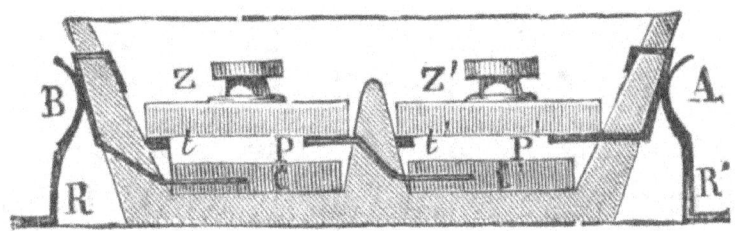

Fig. 57a.

the level of the ledges, and the mixture stirred with a small glass rod, or a bit of stick, to facilitate solution. The zincs are then replaced, care being taken that there be sufficient solution to reach to their under surfaces, when the battery is ready for action. With these small cells, while the E.M.F. is as before about 1·5 or as a maximum 1·52 volts (therefore about three volts for the pair), the internal resistance is rather high, say from 0·75 to 1 ohm. The resulting current is however sufficient for the low resistance of small pocket coils.

§ 73. *Schanschieff's* cell differs from the Marié Davy partly in the arrangement of the elements, but more in

the nature of the mercury sulphate employed.   As .to the
arrangement, it is identical with that of the Walker cell,
illustrated in our Fig. 42, *q. v.*, and consists in a zinc plate
sandwiched between two graphite plates.   The excitant,
for the preparation of which Schanschieff took out a patent,
is made as follows: 3 lbs. of sulphuric acid, sp. g. 1·845,
are added to 2 lbs. of metallic mercury, heated and kept
boiling until the metal is dissolved, and the heat maintained until the excess of acid is evaporated.   When cold,
one gallon of water is added, which only dissolves a part,
precipitation of *basic* sulphate of mercury taking place.
The solid residue is separated and boiled with sulphuric
acid as before, in the proportion of two parts of the residue
to three parts of acid.   By boiling, this residue is dissolved,
and by continuing the heat the excess of acid is driven off.
The .resulting mass when cold is added to the original
solution, which dissolves a part, again separating the
residue, which is treated as before.   After three or four
repetitions, the whole will be found to dissolve in the
original gallon of water.   The density of the liquid will
be 1·435 Beaumé, and the quantity about five quarts.
Finally it is evaporated by heat until the salt is deposited
in a solid state.   It is raked out as it falls, and is packed
in closely-stoppered bottles, in which it may be kept for
an indefinite time.   When liquid is required for use in a
battery, one gallon of water is added to 5 lbs. of the salt,
which dissolves perfectly, leaving no residue, or if there
are impurities, which render the liquid turbid, it may be
filtered through paper.   In another process metallic
mercury is dissolved in sulphuric acid and the excess of

acid evaporated. When cold the solution is dissolved in three times its weight of water, partial solution only, accompanied by precipitation, taking place. Strong sulphuric acid is then added, little by little, and finally drop by drop, with constant agitation; the solution gradually becoming more and more complete, until at last it is perceived that the drop of sulphuric acid, as it falls into the solution, gives rise to a precipitate. At this

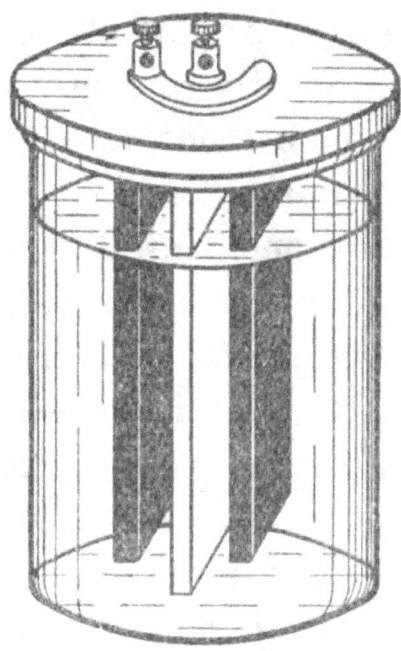

Fig. 58.

point the process is stopped, and the solution allowed to cool, and filtered. It is then evaporated, and yields in a dry state the novel saline material.[1] However the salt be prepared, in order to fit it for use in the battery it is dissolved in water, in the proportion of 1 part of this

---

[1] Since no definite salt corresponding to an *acid* sulphate of mercury is known, it would appear that Schanschieff's salt is simply a mixture of mercuric sulphate with adhering sulphuric acid.

mercury sulphate to 3 parts of water, by weight. The elements are then plunged into the solution, when the cell is ready for use. The current yielded is very steady; and were it not for the troublesome preparation of the mercuric salt, would be far preferable in use to many more vaunted cells. In making up a cell of this kind, it will be found advantageous to have the zinc plate long, so as to reach nearly to the bottom of the cell; while the two carbon plates should be somewhat shorter, as shown at Fig. 58. By this means the mercury which is liberated detaches itself from the carbons, falls to the bottom of the cell, and is absorbed by the zinc.

§ 74. *Gates*, of Bayonne, New Jersey, in 1887 patented another form of single-fluid mercurous sulphate cell, which has several interesting points. The main body of the case or vessel in which the cell is mounted is a horizontal hollow cylinder, with one flat end. Another flat end is fitted so as to extend a little way into the interior of the main tube, so as to form an absolutely tight-fitting plug thereto. The elements (carbon and zinc) are bedded in pitch, and held by screws, which are kept out of contact with the exciting fluid by a covering of pitch. A screw holds one element against one end, and a corresponding screw retains the element in place at the other end. The conducting wires are attached to the screws. When the removable plug is in its place, the only orifice which is open to the atmosphere is a small one at about mid length of the tube, which opens out into a bell-shaped funnel. Fig. 59 is a central vertical section, cut lengthwise; Fig. 60 is a vertical transverse section, while Fig. 61

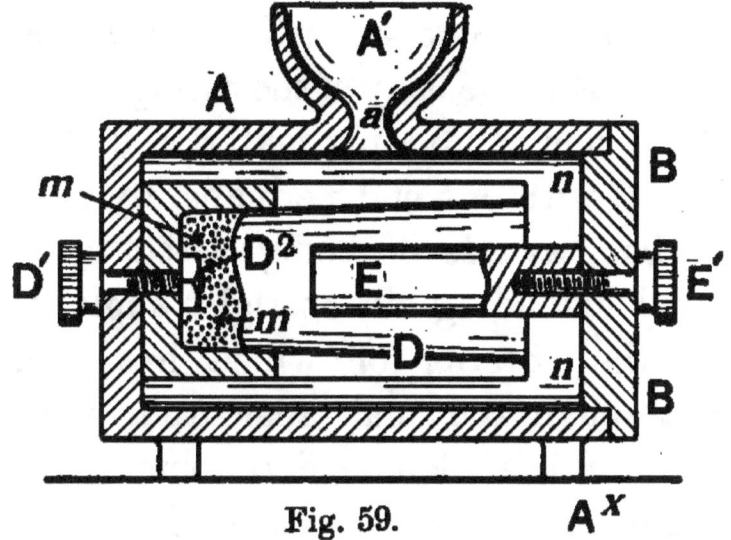

Fig. 59.

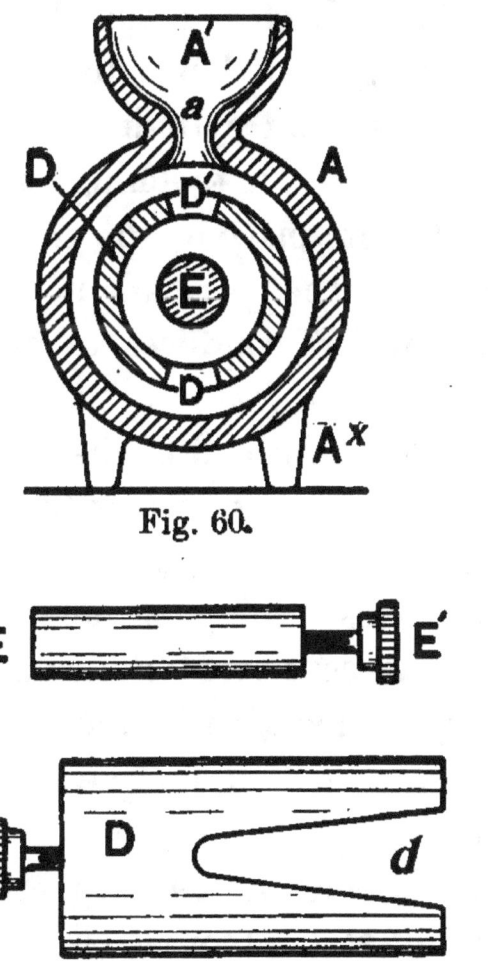

Fig. 60.

Fig. 61.

shows certain parts detached. (The figures are lettered homologously.) *A* is the body of the casing, made of ebonite or similar hard non-conducting material, in the form of a hollow cylinder, with one end permanently closed, and the other open, but capable of being closed by the plug *B*. *A'* is a funnel extending out at right angles to the main cylinder *A*, the interior of which has access to the atmosphere through the small aperture *a*, at the junction of *A* with *A'*. *B* is a plane end or plug with a chamfer turned at its inner extremity, so as to fit accurately in the open end of *A* and make a tight joint all round. *D* is the carbon, *E* the zinc. Each is made of the form represented, the carbon *D* being a hollow cylinder, deeply notched as shown, while *E* is a solid cylinder of smaller diameter, so as to enter into *D* without touching it at any point. The carbon *D* is held in its place by the thumbscrew *D'* inserted in a hole through the permanantly closed end of the cylinder. The nut D^2 conduces to make a good contact with the carbon. Pitch (*m*) is applied between the sides of the carbon and the lower end of the cylinder, also over the end of the nut D^2. The zinc *E* is held in position by the thumbscrew *E'*. Paraffin (*n*) is inserted between the end of *E* and the corresponding surface of *B*. The positive conducting wire is connected by being wound round the screw *D'*, while the negative is wound round the screw *E'*. The lugs A^x are formed in one piece with the cylindrical body *A*. The charge, which consists of a mixture of 1 part by weight of mercurous sulphate ($Hg_2 SO_4$) and 2 parts by weight of sal ammoniac ($NH_4 CL$) powdered and mixed together, is introduced into the cylinder in the dry form. To set the cell in action it is only necessary to pour water in the funnel A'. By keeping the aperture a small, the cell may be inverted or shaken without much risk of spilling, the fluid contents being retained by

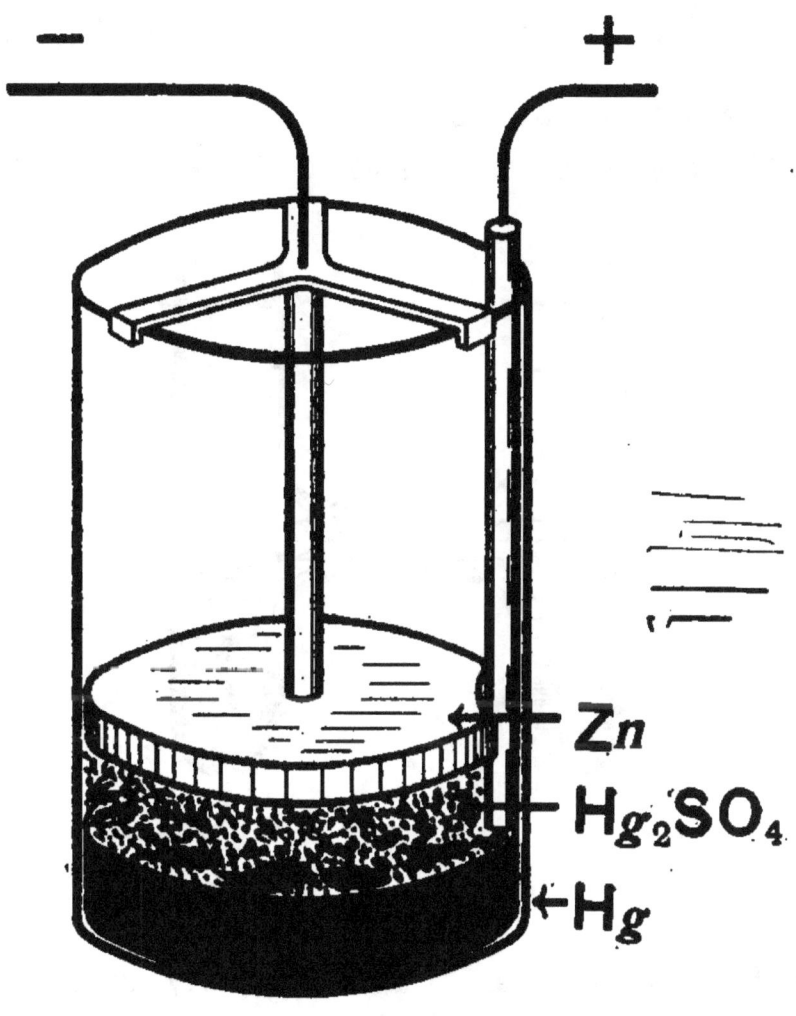

Fig. 62.

capillarity and atmospheric pressure. In putting the carbon D in place, care must be taken that the notches come in a line with the small aperture a, so that the powdered charge introduced may fall in between the elements. This precaution also secures free exit for any gas that may be liberated, and this conduces to de-polarization. The reactions that take place are as follows :—

$$Zn + 2\,NH_4\,Cl, = Zn\,Cl_2 + 2\,NH_4:\ \text{then}\ 2\,NH_4 + Hg_2$$

$$SO_4 = \left.\begin{matrix}NH_4\\NH_4\end{matrix}\right\}SO_4 + Hg_2.\ \text{It is evident that Schanschieff's}$$

salt will answer equally well.

§ 75. Closely allied to these cells we have *Latimer-*

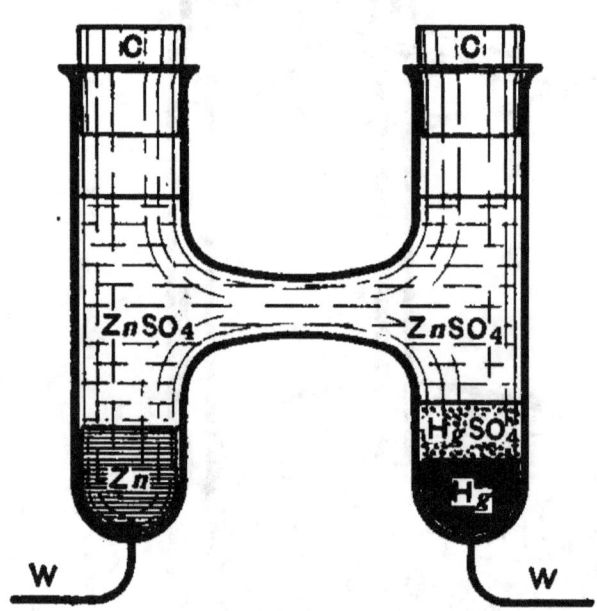

Fig. 63.

Clarke's " standard " cell.   As originally constructed, it consisted in a layer of metallic mercury placed at the bottom of a glass or glazed porcelain jar.   This formed the negative element, and from it rose, to the outside of the vessel, a well-insulated copper wire, as one electrode. Over the surface of the mercury was placed a paste made by boiling mercurous sulphate in a saturated solution of zinc sulphate.   A stout disc or plate of zinc, forming the other element, was placed resting on this paste. A second insulated wire connected to the zinc formed the other electrode. This form is illustrated in our Fig. 62. The E.M.F. at 15° Centigrade is 1·438 (Sloane), or according to Ganot 1·495. The E.M.F. *diminishes* as the temperature rises, at the rate of about ·00078 volt for each degree Centigrade. There are several modifications of the Clarke cell, of which the following are the more important—(*a*) the "*H*" type. In this, as shown in Fig. 63, we have two tubes closed at the bottom, in which are fused two platinum wires, *W W*, which serve as electrodes. At about the centre of the tubes there is a cross tube connecting the two together. In the bottom of the left-hand tube, and in contact with *its* platinum wire, is placed an amalgam of pure zinc, Zn, while the right-hand tube contains a layer of pure mercury, Hg, covered with mercurous sulphate, $Hg_2 SO_4$. Both tubes are then filled, to just above the level of the cross tube, with a saturated solution of zinc sulphate, S, to which a few crystals of the same have been added. The tubes are then corked at *C* to prevent loss of fluid by evaporation. A diminution in the strength of the zinc sulphate solution increases the E.M.F. This cell polarizes rapidly, and it is only fit for use as a standard of E.M.F., for which purpose, however, its temperature coefficient is considered too high. (*b*) *Lord Rayleigh's* modification, also known as Clarke's "Normal" cell, is represented at Fig. 64. Here we have but one tube, with a single platinum wire sealed in the bottom, this wire being in contact with the layer of mercury, Hg. The top of the tube is fitted with a stopper from which hangs a short rod of pure zinc attached to a wire passing through the stopper.   Over the mercury, and reaching to

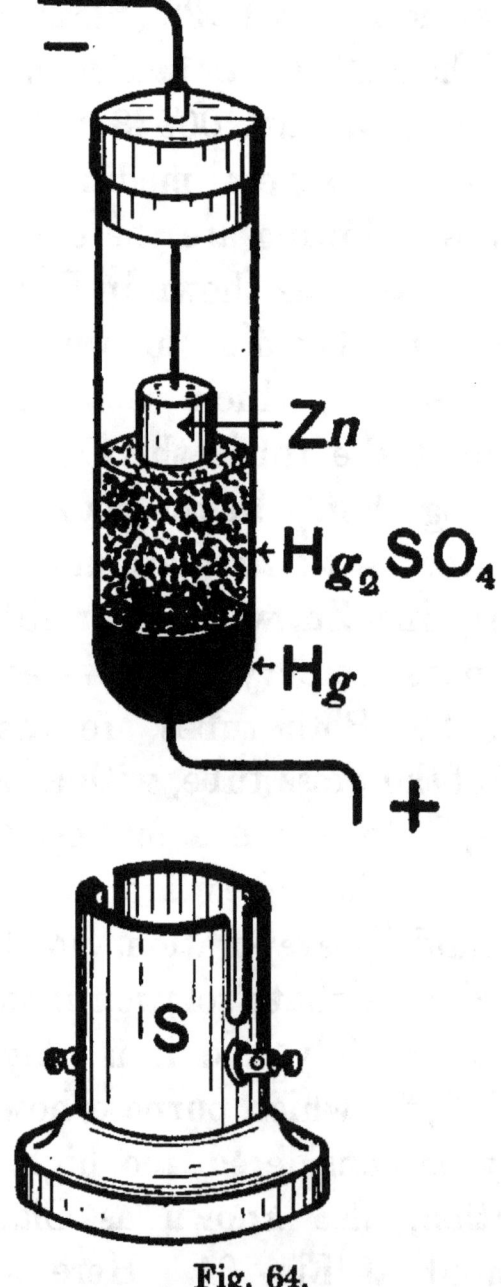

Fig. 64.

the zinc, is placed a thick semi-fluid paste made of mercurous sulphate and solution of zinc sulphate. For convenience of manipulation, this form is generally provided with a little cleft wooden stand S fitted with two terminals, in which the two wires $+$ and $-$ can be inserted to facilitate connection to the outer circuit. According to Lord Rayleigh, the E.M.F. of such a cell is in true volts at 15° Cent. 1·435, with a temperature coefficient of $1 - ·00077 (t - 15)$, when t is the temperature in degrees Centigrade. (c) Another modification of the Clarke cell, due to *Henry S. Carhart* (1895), with a view to portability, consists in casting the zinc with a foot, which nearly fits the tube; this foot holds down some purified asbestos that retains the mercury and mercurous sulphate in place during transportation. The stem of zinc is enclosed in a glass tube, cemented on with a non-conducting wax. Fine crystals of zinc sulphate are mixed with the mercurous salt, and more are placed on the top of the zinc plate. Thus the zinc is only in contact with liquid at the bottom.

§ 76. Nearly related to the mercurous sulphate cells just described are those in which lead sulphate or lead chloride is employed as a depolarizer. Of these the most important are :—

§ 77. *Becquerel's* lead sulphate cell, in which the negative element was a lead, copper or tin rod, around which was cast a cylinder of sulphate of lead. Becquerel recommended the addition of $\frac{1}{8}$ common salt to the sulphate previous to fusing and casting. A zinc cylinder or a zinc rod formed the positive element. The excitant may be dilute sulphuric acid, strength 1 part acid to 12 of water. The E.M.F. of this cell is only 0·5 volt, while its internal resistance is very high. It has never come into very extensive use. This cell may be also put up precisely like the Marié Davy, using a carbon as the under plate, with the lead sulphate sprinkled over it, on which is placed the dilute acid, surmounted by the zinc, supported at a little distance from the lead sulphate. It is doubtful whether, in view of the very imperfect solubility of the lead sulphate, it conduces much to the constancy of the cell. In pure water it is practically insoluble; in dilute mineral acids only slightly so. It is readily dissolved by caustic alkalies, and in many ammoniacal salts, also in sodium thiosulphite (hyposulphite of soda), and it is probable that with these solvents the Becquerel cell would be more efficient.

§ 78. LEAD CHLORIDE has also been used as a depolarizer in a cell similar to the above, the only difference being that dilute hydrochloric acid is then employed instead of sulphuric acid as the excitant.

§ 79. Wheatstone, De la Rue, Niaudet and others have employed the higher oxides of lead, such as lead peroxide, $Pb\, O_2$, and minium (red lead), $2\, Pb\, O + Pb\, O_2$, as depolarizers; and as these oxides are fairly good conductors of electricity, they may be employed directly as the negative elements. Several good forms have been employed, of which we will notice the following, viz. :—

§ 80. The *Emile Reynier* cell (1884), in which an amalgamated zinc plate is opposed to one or two lead plates, on which lead peroxide had previously been deposited electrolytically, or else pasted on mechanically. The excitant used is dilute sulphuric acid, sp. g. about

1·15 (about 1 pint strong acid to six of water). The E.M.F. of such a cell is about 2·36 volts, and its internal resistance very low, so that a large current can be obtained from it. Its inventor showed its capabilities as a storage cell by "forming" the lead peroxide on the surface of the lead plate by means of the dynamo or battery current, while in the acid solution. He constructed a special form

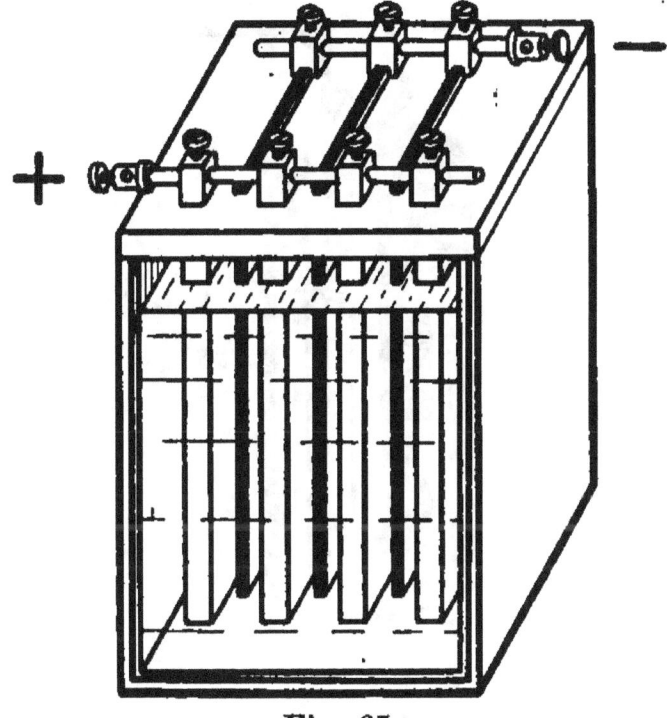

Fig. 65.

of battery, of which we give an illustration at Fig. 65, for this particular purpose. To prevent any chance electrical leakage the case of the cell is jacketed; and between the inner and outer case, pitch, paraffin wax, or other good insulator, is run in. The three thinner plates are of well-amalgamated zinc;[1] the stout plates are made of lead, on

---

[1] When used as a "storage" cell, the zinc was electrolytically deposited on the surface of a thin lead plate.

M

which lead peroxide is deposited by electrolysis. Such a cell, with the three zincs joined together to form one element, and the four lead connected as shown, to form the other, the total superficial area of this latter being 200 decimètres (about 3136 square inches) will give an E.M.F. of 2·36 volts, and will have an internal resistance of about 0·02 ohm; in other words, it could give as much as 118 ampères on the short circuit. As a matter of fact, it

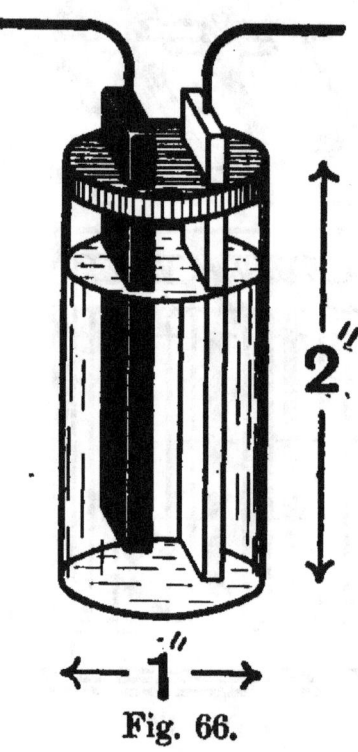

Fig. 66.

would never be advisable to draw off a larger current than 25 ampères from such a cell, since any greater flow would heat the cell and produce injurious disintegration of the lead peroxide plates. But 25 ampères is a fair rate of discharge for such a cell, so that any external resistance may be used, provided it is not less than ·0744 of an ohm.

§ 81. The "*Kingsland*"[1] cell (Fig. 66) is very similar to

[1] First described in 1890.

the Reynier cell just described, except that the peroxide of lead element, instead of being electrolytically prepared, consists in that block form known as " Lithanode." Lithanode (" Stony Anode ") is a composition devised by Mr. Desmond Fitzgerald, and described by him in 1886 as a mixture of oxides of lead with glycerine, sulphate of ammonia, or dilute sulphuric acid, which is placed in moulds of the shape and size desired to be given to the plates, the mass being then subjected to great pressure in a hydraulic press. The compressed plates are then transferred to a drying room; and when perfectly set and hardened, they are immersed into a solution of hypochlorite of magnesium (or even of chloride of lime) which converts the surface into lead peroxide. Finally, the peroxidization of the entire substance of the plate is carried to the highest possible point, by electrolysis, in a bath of sulphate of magnesia. Some of the best plates contain as much as 90 per cent. of pure peroxide. The lithanode plate thus prepared, is said to be the only one which does not disintegrate when placed in a fluid. When used in the " Kingsland " cell, against a zinc plate acted on by dilute sulphuric acid, sp. g. 1·170, the E.M.F. is 2·35 volts, and this remains constant for a very considerable time, if there be a fairly high resistance in circuit. As the cell is sealed, this form is very convenient for testing purposes, more especially as when it has run down it can easily be re-charged (like a secondary cell), by passing a small current (about 0·5 ampère) through it, from the lithanode plate to the zinc.

§ ·82. In order to make use of broken pieces of lithanode, the author devised a modification of the above cell, which gave very good results. A cylinder of well-amalgamated zinc is inserted into a glass or stoneware jar; a large rod of carbon is fitted with a flannel bag, reaching nearly to the zinc cylinder. This bag is filled up to the level of the top of the jar, with the broken pieces of lithanode, which must be pressed down, so as to make good contact with

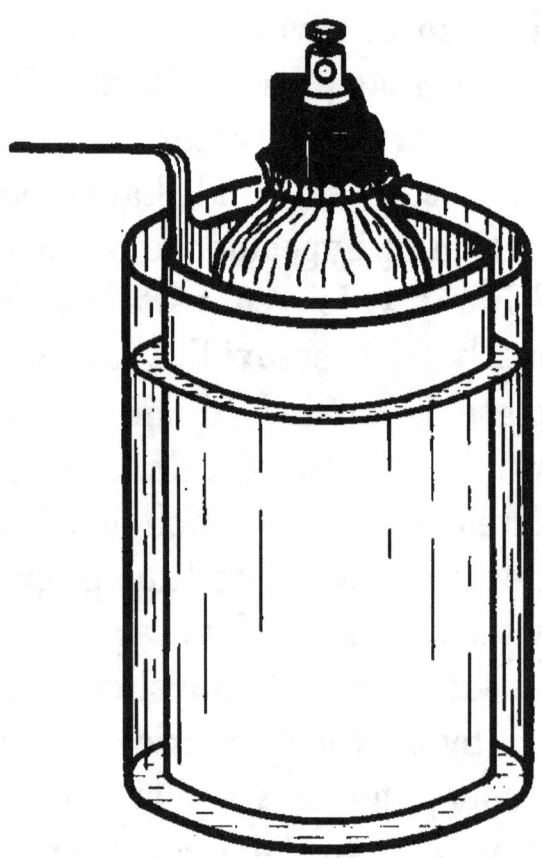

Fig. 66a.

the carbon rod. Terminals are attached both to the zinc plate and to the central carbon. The jar is then filled with dilute sulphuric acid, sp. g. 1·15. Owing to the imperfect contact of the lithanode pieces, the resistance is rather higher than that of a cell in which plates are employed instead of pieces; but the difference is not very great. The reaction which takes place in these lead peroxide cells is represented below :—

$$Zn + H_2 SO_4 = Zn\ SO_4 + H_2;\ \text{then}\ H_2 + Pb\ O_2 = H_2 O + Pb\ O;$$

and if the action be continued long enough, the Pb O is reduced to metallic lead, in a spongy condition, by the advent of fresh hydrogen. This cell is illustrated at Fig. 66a.

§ 83. De la Rue was, we believe, the first to use *manganese dioxide* (Mn O_2) as a depolarizer. It is a powerful oxidant, and has had extensive use in cells of the Leclanché type, which will be mentioned among the two fluid cells. Here we need only notice the *Agglomerate* or *Block Leclanché*. This consists of a well-amalgamated zinc rod or plate, held in proximity to one or more blocks (made of a compressed paste of graphite and manganese dioxide) by means of suitable india-rubber bands, of such shape, as to prevent the agglomerate blocks from actually coming into contact with the zinc. A wire is cast in the zinc to form the negative terminal, a binding screw being clamped to a central carbon plate, to form the positive terminal; or if more blocks than one are used, these are united together at the top by an india-rubber band, so as to form but one element. The exciting solution is a half saturate solution of ammonium chloride (sal ammoniac) $NH_4\ Cl$.[1] The paste with which the agglomerate block is made, consists in a mixture of 40 parts manganese dioxide, 52 parts of carbon, 5 parts of gumlac, and 3

---

[1] About 4 oz. to the quart of water.

parts of potassium bisulphate. This mixture is placed round a carbon (graphite) core, of suitable shape and size, and forced into a steel mould, at a temperature of 212° Fahr., and then subjected to a pressure of about 4500 lbs. to the square inch. The blocks are then subjected in a closed oven to a temperature of about 662° Fahr., which

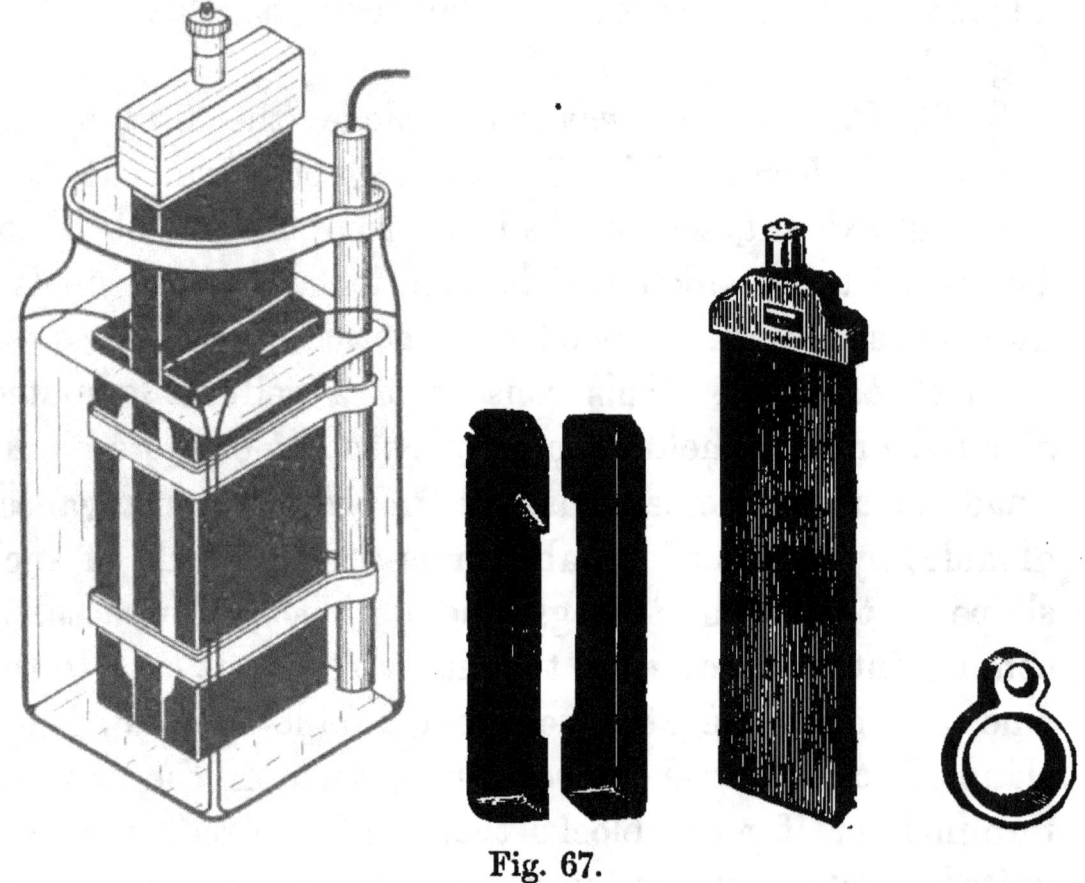

Fig. 67.

is insufficient to eliminate the oxygen from the manganese dioxide, but which drives off all moisture and volative matters, and causes the gumlac to flow and agglomerate the other materials, so as to render the block solid enough to resist disintegration when immersed in the fluid contained in the cell. The purpose of addition of the potassic bisulphate to the paste, is to render the resulting block porous to some extent, so that the fluid can enter into these pores, and thus allow the liberated hydrogen to come into contact with the manganese dioxide, thereby ensuring depolarization. The Agglomerate cell takes two forms, the "double block" and the "six block." In the former we have a central carbon plate (see Fig. 67) with an

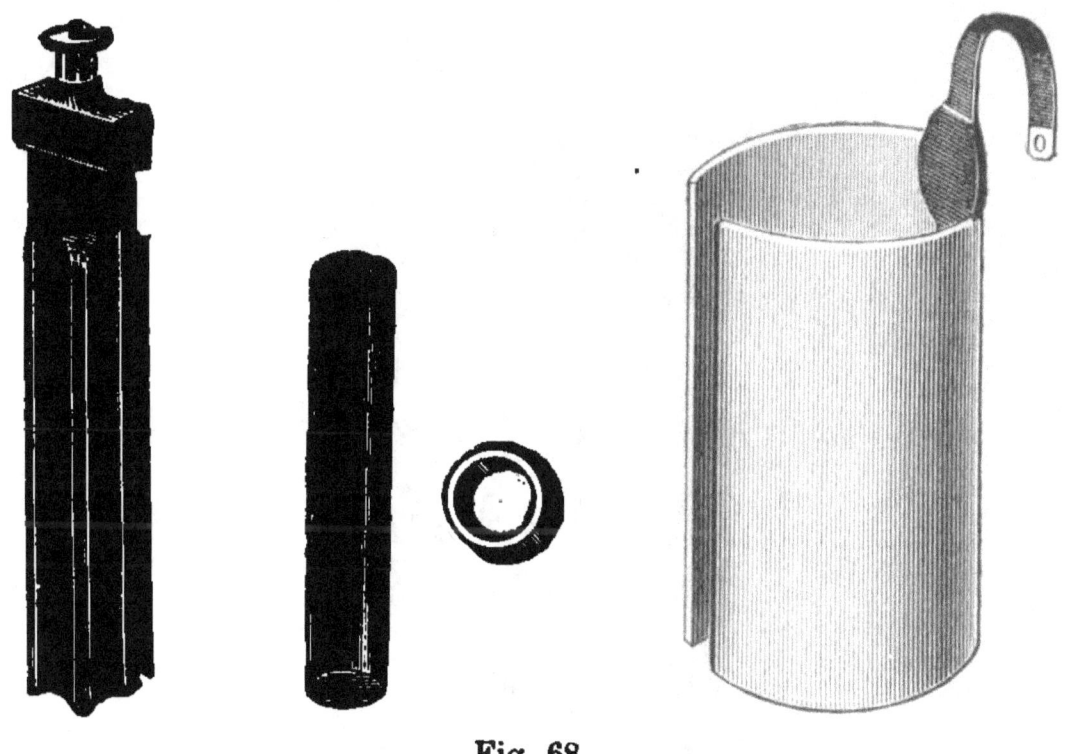

Fig. 68.

agglomerate plate on each side, both of which are kept closely in contact with the carbon by means of an elastic band; this band has a loop on one side, through which the zinc rod passes. In the six-block form, Fig. 68, the carbon is made in the shape of a rod with six deep channels in the sides. These just allow six blocks of the agglomerate material to lie in them, and the whole is held firmly together by two elastic bands, one at the lower and the other at the upper extremity. The zinc takes the form of a hollow cylinder, and is furnished with an extension or "lug," whereby connection can be made to the outer circuit. The complete cell is shown at Fig. 69. Manganese dioxide batteries are much used where small

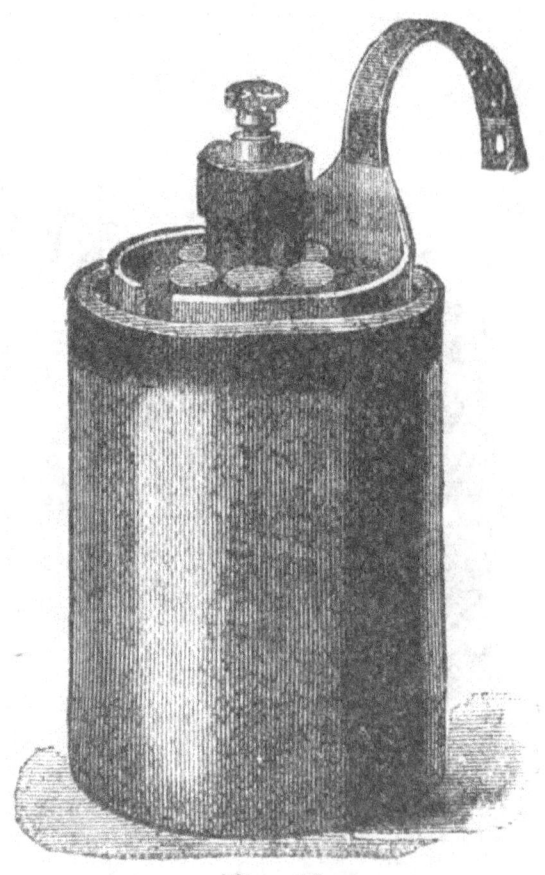

Fig. 69.

currents are required for short runs at intervals extending altogether over considerable periods, say a year or more; as, for instance, working bells, relays, electric clocks, etc. The E.M.F. is nearly 1·5 volts; the internal resistance of a cell of the capacity of 3 pints, is about 0·36 ohm. The reaction that takes place is as follows :—

$$Zn + 2\,NH_4\,Cl = Zn\,Cl_2 + 2\,NH_3 + 2\,H$$

Zinc and ammonium give zinc plus ammonia and free hydrogen.
chloride chloride

In the presence of the manganese dioxide, the hydrogen seizes upon its oxygen to produce water, while manganese sesquioxide is left, thus :—

$$2\,Mn\,O_2 + 2\,H = Mn_2\,O_3 + H_2O$$

Manganese and hydrogen give sesquioxide and water.
dioxide of manganese

While manganese dioxide is a fairly good conductor of electricity, the sesquioxide is but a poor one, so that as the work goes on, the resistance of the cell increases, with a corresponding decrease in the volume of the current available. The manganese dioxide can however be re-generated, by soaking the blocks in a strong solution of any of the permanganates (potash, soda, or ammonium salts), such as Condy's fluid, as long as such immersion de-colorizes the liquid. It is however rather doubtful economy. It will be noticed that ammonia is given off during the action of the battery, and this attacks all the brass work, terminals, etc., producing therewith verdigris, and thus interfering with the goodness of the contacts. If sufficient sal ammoniac be not present, the zinc acts in the water alone, giving rise to zinc oxide, which forms white incrustations on the zinc, and renders the fluid milky. If this should be noticed, it will be advisable to add more sal ammoniac to the solution. For every 100 parts of zinc consumed in such a cell about 164 parts of sal ammoniac are required, and the liberated hydrogen requires 248 parts of pure manganese dioxide to convert it into water. Ordinary commercial " black oxide of manganese" is not pure dioxide, so that considerably more must be allowed in making up the cell. Owing to the slowness with which the hydrogen reacts on the manganese dioxide, it is not possible to get a steady current from such a cell for more than a few minutes at a time. Hence these cells are of no use for long or heavy discharges, such as for working coils, lighting lamps, or driving motors. But a short interval of repose enables the reaction to take place, when the cell soon recovers its activity. As little or no action takes place when the circuit is broken, no waste of zinc occurs, and as no noxious fumes are given off, these manganese dioxide cells are extremely handy for occasional use, more especially as they require no attention beyond the addition of water, to make up for loss by evaporation. All cells containing saline solutions are subject to the evil of "creeping." By creeping is understood the gradual crystallization of the salts on the inside of the containing vessel, *above* the level of the contained fluid. Little by little this crystallization extends up to the top of the jar, and even "creeps" over to the outside, and this tends to short-circuit the cell. But this inconvenience can be easily guarded against, by dipping the containing vessel (before making up the cell), mouth downwards into melted paraffin wax, as far as the exciting fluid is intended to reach. This effectually prevents the formation of crystals, and thus puts a stop to the creeping.

§ 84. In *Dr. Conrad Pabst's* cell we have oxide of zinc used as the depolarizer. In general construction it is very similar to the Agglomerate just described. The negative element consists of a carbon block "prepared with oxide of zinc," as stated in the patent dated 1887. The positive element is a sheet or rod of zinc. Both elements are suspended from a cross piece resting on the edge of the top of the glass containing jar, and they dip into a solution of zinc chloride. The E.M.F. of this cell is rather over 1 volt, and the internal resistance about 1 ohm in the quart size cell. It is said to be a fairly constant cell, with good lasting power, provided too heavy a current be not taken from it. It has been pretty extensively used on the German State Railways, and by the German Telegraph Administration. The cell is interesting from a theoretical point of view, as showing practically the effect of the "migration of the ions" (see § 31 *et seq.*). One would hardly think that a solution of chloride of zinc would act on metallic zinc; and in fact it does not, unless an electro-negative body be placed in opposition to it and be connected metallically to it, so as to allow the difference of electrical condition to manifest itself. Given these conditions, certain exchanges in the relative positions of the ions take place, as shown by the following equations. The cell originally contains a solution of zinc chloride $Zn\,Cl_2$, in water H_2O. When the zinc plate or rod is connected up to the carbon the changes symbolized below occur :—

$$Zn\ +\ H_2O\ =\ Zn\,O\ +\ H_2$$

Zinc and water give zinc oxide and free hydrogen.

This free hydrogen then attacks the zinc chloride, giving rise to hydrochloric acid and free zinc,

$$H_2\ +\ Zn\,Cl_2\ =\ 2\,H\,Cl\ +\ Zn$$

Hydrogen and zinc chloride give hydrochloric and zinc.
acid

From molecule to molecule, in the chain reaching from the zinc plate to the zinc-oxide incrusted carbon, these interchanges go on; here the free hydrogen seizes on the oxygen of the zinc oxide to form water, which mixes with the solution to undergo again the same cycle of changes, while the liberated zinc is taken up by the freshly-formed hydrochloric acid. But when the circuit is opened, and connection between the zinc and carbon interrupted, all action ceases.

§ 85. To *De la Rue* we are indebted for the application of silver chloride as a depolarizer. In the original form of this cell, a stout silver wire formed the negative element. Around this was cast a cylinder of silver chloride, Ag Cl,[1] and this was placed in a small cell, along with an unamalgamated zinc rod. The excitant employed was a solution of common salt Na Cl. The zinc and salt reacted, giving rise to zinc chloride $Zn Cl_2$ and sodium Na. In the presence of water this latter instantaneously seized upon the oxygen of the water, with the formation of soda, and the liberation of hydrogen. This hydrogen in its turn seized upon the chlorine of the silver chloride, producing metallic silver, and hydrochloric acid, thus $H_2 + 2 Ag Cl = Ag_2 + 2 H Cl$. The silver thus liberated, thickened the original silver wire; the hydrochloric acid produced, combined with the soda first liberated, to form salt and water, thus $2 H Cl + Na_2O = 2 Na Cl + H_2O$. A modification consists in using a glass tube 6″ long by ¾″ diameter inside, closed by a vulcanized india-rubber stopper,

[1] The precipitated salt is not so active; it must be *fused* to give good results.

which is pierced in two places—one hole being central, to allow the passage of a flattened silver wire $\frac{1}{16}''$ broad and 8″ long, reaching to the bottom of the tube; the other hole more to the side, which admits of a well-amalgamated zinc rod $4\frac{1}{2}''$ long $\frac{3}{16}''$ diameter. To prevent the silver being

Fig. 70.

affected by the sulphur in the vulcanized rubber stopper, it is wrapped round for about half its upper length with thin sheet gutta-percha (see Fig. 70). In the bottom of the tube are placed 225·23 grains, about half-an-ounce, or 14·59 grammes of powdered silver chloride. This is the depolarizer. Above this is placed a solution of sodium chloride, containing 25 grammes in 1 litre of water, or 1752 grains in 1 gallon, or a solution of ammonium chloride, containing 23 parts in 1000 parts of water, which is the approved excitant. This liquid is poured in until the level rises to about 1 inch from the stopper. The silver is surrounded by a vegetable parchment tube, through which it passes at the top, to separate it from accidental contact with the zinc rod. Water can be added through a plugged hole in the stopper. Connection of one cell to another is made by passing a short piece of india-rubber tube over the zinc of one cell and drawing the silver wire of the next cell in so as to press against the zinc. E.M.F. about 1·3 volts with silver chloride, or ·908 with silver bromide, or ·758 volt with silver iodide as the depolarizer, or of old form with ordinary salt, ·97 volt. The internal resistance of the modified form, with ammonium chloride, 4·2 ohms. The cell used by De la Rue and Müller in 1868 was suggested by Marié Davy in 1860. A new cell gives a very feeble current at first, but after a time the reduced silver of the silver chloride forms a larger silver surface, and the current attains its normal strength. Warren de la Rue and Müller used a battery of 14,400 such cells for their researches on the electric discharge in rarefied gases. A battery of 8040 cells gave a spark $\frac{1}{3}$ of an inch long in air at the ordinary pressure, and $1\frac{1}{5}$ inch under a pressure of a quarter of an atmosphere. The length of the arc found between the terminals varies with the square of the number of cells in series. Thus, while 1000 cells give a spark of ·0051 inch under ordinary atmospheric pressure, 11,000 cells give a spark of ·62 inch. The equation representing the action is probably as follows :—

$$2\,NH_4\,Cl + Zn = Zn\,Cl_2 + 2\,NH_3 + H_2.$$

The liberated ammonia reacts with the zinc chloride :—

$$2\,NH_3 + Zn\,Cl_2 = (NH_2)_2\,Zn\,Cl_2 + H_2.$$

The equation representing depolarization :—

$$4\,H + 4\,Ag\,Cl = 4\,HCl + 4\,Ag.$$

Sometimes the tubes are stopped with paraffin instead of vulcanized rubber. *Pincus* constructed a battery for medical purposes, using silver chloride independently of De la Rue.

§ 86. *Gaiffe's Cell.* Constituents :—Zinc and silver chloride, in zinc chloride. Arrangement :—Outer cell made of ebonite to which a lid is screwed. The two electrodes are fixed to the lid. The negative electrode, consisting of a cylinder of fused silver chloride, is placed in a copper vessel covered with linen. The electrodes are kept in position by india-rubber rings and pieces of caoutchouc. The cell must not be upset, as by wetting the lid the cell becomes short-circuited. Gaiffe prevented this by putting several layers of filter paper, wetted with a solution of zinc chloride, between the electrodes instead of the liquid. E.M.F. 1·02 volts.

Remarks :—Intended for medical purposes, also used as a standard cell with condensers and electrometers. The cylinder of silver chloride may have a silver wire imbedded in it.

§ 87. *Skrivanow's Cell.* Constituents :—Zinc and chloride

of silver in a solution of caustic potash.　Arrangement :—
The cell is similar to De la Rue's, the silver chloride
being in a parchment paper receptacle.　The exciting
solution is 75 parts of potassium hydroxide in 100 parts of
water.　E.M.F. = 1·45 to 1·5 volts.　A cell weighing 100
grammes can give out 1 ampère for 1 hour.　After that

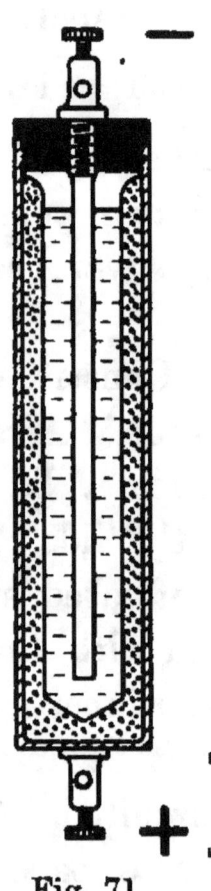

Fig. 71.

amount of work, the potash must be renewed, and also
after two or three changes of potash the silver chloride
needs replacing.

§ 88. *J. Mackenzie*, in 1882, patented a modification
of the De la Rue cell.　With the intention of rendering
the cell less liable to fracture, and therefore portable as

a pocket cell, he constructed the outer case in the form of a copper tube, fitted with a copper plug at the bottom. This was heavily silvered in the inside. The interior was then coated with a layer of fused silver chloride. A stopper of ebonite, or other suitable insulator, impermeable by moisture, was made to screw into the top of the tube at the open end, and through this stopper passes a zinc rod or cylinder, having a screw thread cut at its upper extremity, so as to make a watertight fit. This rod of zinc reaches nearly to the bottom of the copper cylinder. The exciting fluid consists of an aqueous solution of either zinc chloride, zinc sulphate, or sodium chloride, preferably the former. Fig. 71 is a sectional view of the Mackenzie cell.

§ 89. Hitherto we have considered cells in which the excitant has been either a dilute acid (a hydrogen salt), or else a saline solution. We now pass to study certain cells in which the excitant is an alkaline solution, such as, for instance, a solution of caustic potash, caustic soda, or caustic ammonia in water. One advantage in such excitants is, that practically no action takes place between the zinc (or other positive metal) and the excitant, unless the circuit be closed. Hence little or no waste of zinc occurs on the open circuit. Another good point about them is, that being very good conductors, there is but little internal resistance in cells in which they are employed. The chief drawback lies in the fact that the E.M.F. is small (rarely exceeding 0·8 volt per cell). The first of this class to demand our attention is the *Lalande* cell (1882). The constituents of this cell, as originally devised by Lalande, are a plate or rod of zinc, as the positive element, a 30 per cent. solution of caustic potash in water as the excitant, and a plate or box of iron or copper as the negative element, in contact with black oxide of copper (cupric oxide: CuO) as the depolarizer. The original arrangement consisted in a glass vessel with a tightly-fitting copper lid, held in place by a wide india-rubber band. From the inside of the lid depended two sheet-iron or sheet-copper plates, attached thereto by screws. On the inner surfaces of these plates were fastened two blocks of compressed cupric oxide,[1] by means of india-rubber rings sprung on. A zinc rod passed through a hole in the centre of the lid; and to prevent contact with the same, it was encircled at its upper extremity by a glass tube, caused to fit in the hole by an india-rubber washer. As gas is generated during the action of the battery, a second glass tube was passed through the copper lid: this exit being closed by a rubber valve, which opened if the pressure of the gas accumulating in the vessel became too great (see Fig. 72).

§ 90. The next modification was *Lalande's Trough Cell.* The constituents were zinc, cupric oxide, and sheet-iron. The following was the arrangement :—The sheet-iron was made into a waterproof trough 16 inches long, 8 inches wide, 4 inches high. The bottom of this

---

[1] These copper blocks may be formed by mixing from 5 to 10 parts of a mixture of equal parts of magnesium chloride and magnesium oxide, with from 95 to 90 parts of black oxide of copper, with sufficient water to form a stiff paste, which is compressed into mould of the desired shape, allowed to dry, and then gently heated, not so much as to drive off the oxygen. Such blocks are fairly porous.

was covered with a layer of oxidized copper (not compressed). At the four corners, porcelain insulators were placed to carry the horizontal zinc plate, one side of which was bent up at right angles, for connection to the outer circuit by means of a terminal. Another terminal was fixed to the iron trough. . The iron vessel was about

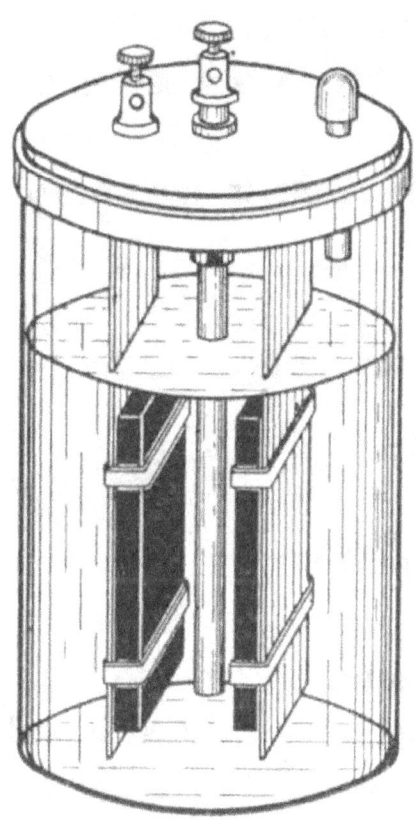

Fig. 72.

three-quarters filled with caustic potash solution, containing about 30 to 40 parts of caustic potash in 100 parts of water. To prevent the combination of the potash with the carbon dioxide of the air (atmosphere), and also to minimize evaporation, either the cell was closed with a lid, or a layer of petroleum was poured over the caustic

potash solution.   According to experiments performed by
E. Hospitalier, published in *L'Électricien*, 1883, the
Lalande cell is superior to the Daniell in using less zinc,
and preferable to the Leclanché in being more constant.
For every gramme of zinc consumed nearly 3 grammes
of potassium oxide and 1·35 gramme of copper oxide are
used.   The resistance of the cell decreases during use,
owing to the reduction of the copper oxide to copper.
E.M.F., ·98 volt.   Fig. 73 is a sectional view of the trough
cell.

The *Plate Cell* is another modification of the Lalande.

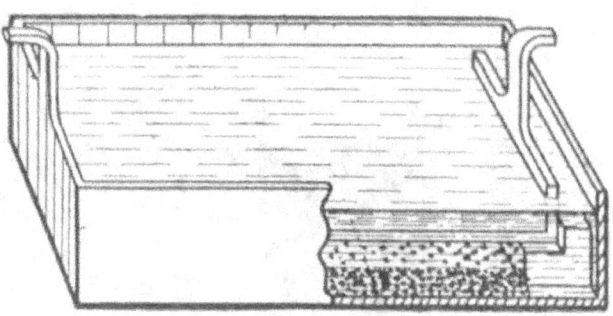

Fig. 73.

Arrangement :—From the ebonite lid of a vessel in which
the zinc rod stands, is suspended a well-insulated copper
wire, that is attached to the bottom of a closed carbon
cylinder.   The inside of the carbon cylinder is filled with
oxide of copper.   The excitant is the same as previously
described.   This cell has the advantage over the Lalande
cell in that when not in use the copper becomes reoxidized.
When the cell is not closed by an outer circuit, the carbon
and copper in the liquid form a closed cell, which gives off
hydrogen at the carbon and oxygen at the copper; the
former escapes, the latter oxidizes the copper.

The *Chaperon Cell.* Arrangement:—A cylindrical glass vessel is fitted with a box made of sheet-iron, which contains the copper oxide. To the box is fastened a well-insulated copper wire. The zinc electrode takes the form of a spiral, and is connected to the terminal at the top of the cell. The straight end of the zinc which does not reach the liquid, is encased in an india-rubber tube. The exciting fluid is, as usual, a solution of 30 to 40 parts of potassium hydroxide in 100 parts of the fluid, the solvent being water. To avoid the handling of the potassium hydroxide, the required amount of oxide is placed in the copper oxide box, and covered with a lid. When the cell is to be used, water is added first, and then the necessary amount of copper oxide.

The following are three modifications, by which all deposition of copper on the zinc is avoided. Whereas the old form was a zinc electrode, as negative pole, and a disc of agglomerate oxide of copper as the positive pole, in a 30 or 40 per cent. solution of caustic potash, the new ones have the zinc arranged differently, also the oxide.

First, *a small-sized cell.* Arrangement:—The zinc is a curved plate, and is hung by a hook from the edge of the glass containing cell, and does not reach below the middle of the solution. Opposite it is placed, also hanging by a hook, a perforated sheet-iron cylinder, containing the copper oxide, and covered with a layer of porous material of very low resistance.

The cell is 8 inches high, 4 inches in diameter, is capable of giving 75 ampère hours, and producing an E.M.F. of

·8 volt, with a normal discharge of 1 ampère, although it will give 2 or 3 ampères on very low resistances.

Secondly, *a medium-sized cell*. Arrangement:—The zinc is a cylinder, hung by a hook as before, and only reaching half-way down to the bottom. In the centre of the enclosed space is a cylinder of copper oxide, resting on the bottom of the vessel.

This cell is 13 inches high and 6 inches in diameter, and will give 300 ampère hours, at a normal rate of 3 to 4 ampères.

Thirdly, *a large cell*. Arrangement:—A zinc cylinder arranged as before, and enclosing a cylinder of copper oxide held at a distance from the zinc by four porcelain cylinders. It is capable of giving 600 ampère hours, at a normal rate of 5 to 6 ampères, and up to 12 to 20 ampères on occasion. Its height is 14 inches, and its diameter 7 inches.

In setting up the cell, the potash is placed in tinned iron boxes and hung from the top, in water. The water enters these boxes by the perforated bottoms, and quickly dissolves the potash, giving rise to a thick solution which sinks to the bottom. The liquid is then mixed, and is ready for operation. This is almost the only primary cell capable of giving a large discharge which does not consume zinc on the open circuit. It has been much used for the electric ignition of gas and oil engines.

§ 90a. *Hartmann's* cell is a flat form of the Lalande-Chaperon. A zinc gauze or perforated sheet of zinc is placed at the bottom of the cell, over which is placed a pad of any porous material, on which is laid a plate of iron or silver. Caustic potash or soda solution is used as the excitant. Not so good as Lalande's, as there is no depolarizer.

§ 91. *Edison-Lalande Cell.* Constituents:—Zinc and copper oxide in caustic potash solution. Arrangement:— The zinc is in the form of a small plate fastened to the porcelain lid by one terminal, and is placed opposite a compressed cake of oxide of copper that is held between

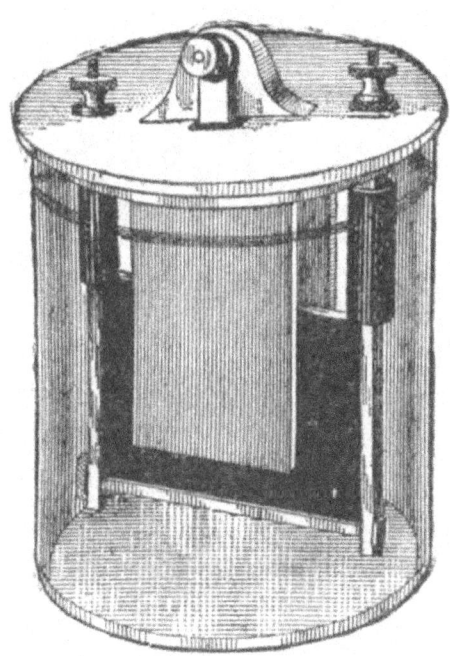

Fig. 74 (*A*).

two strips of copper. These copper strips are clinched together by a screw and nut, so as to grip the oxide cake at its two sides, leaving the faces exposed. Neither plate goes to the bottom of the cell. The E.M.F. of the Edison-Lalande cell is only 0·75 volt.

This low E.M.F. is compensated for by the very small internal resistance, which conjoined to the rapidity with which depolarization takes place, permits of a very large current being taken off, for a long time. Fig. 74 gives

a general view of the interior arrangement of this cell.
In the larger cells of this type (*B*), a single copper oxide
block is placed between two zinc plates. In all cases, to
prevent the potash solution absorbing carbonic acid from

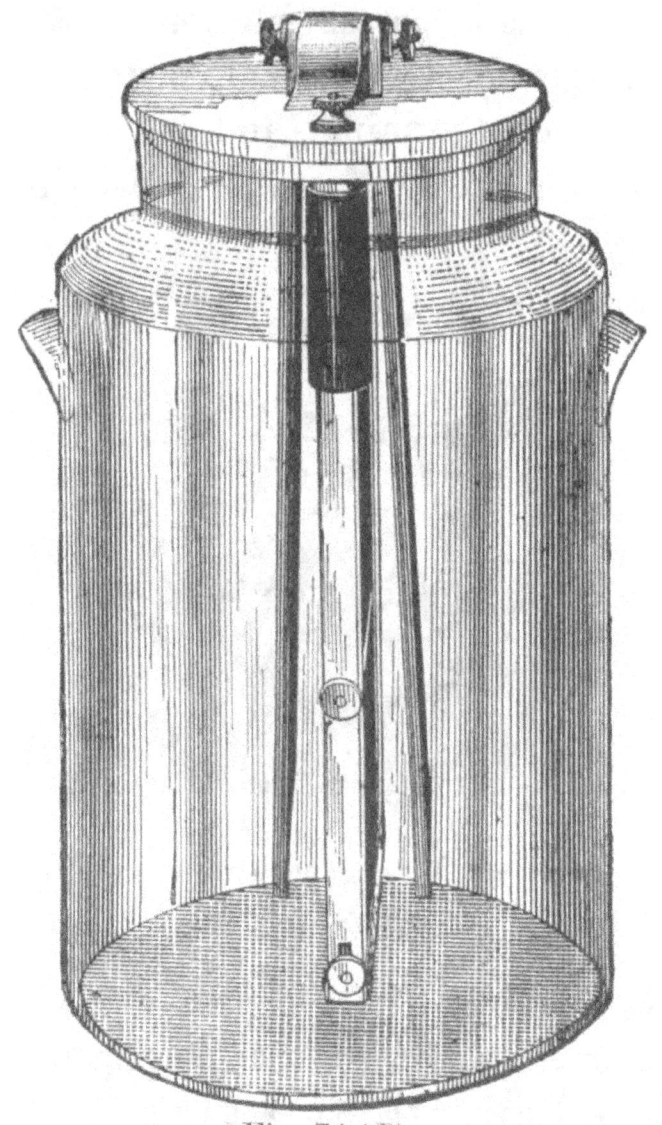

Fig. 74 (*B*).

the atmosphere, a layer of heavy petroleum oil is poured
over the top of the caustic potash solution. Cells of
the smallest size (*A*) (3¾″ diameter, 7″ high) have an
internal resistance of about 0·03 ohm, and a capacity of
about 50 ampère hours; the larger size (*B*), with double

zincs (cell about 7″ diameter, 18″ high) have the same resistance, but a capacity of about 600 ampère hours; in other words, such a cell can supply a current of 1 ampère for 600 hours right off.

§ 92. Under the name of the *Cupron-Element*, Mr. H. Leitner has introduced yet another form of the Lalande cell. The chief difference lies in the mode of the formation of the copper oxide block, which serves at once as the conductor and the depolarizer. We will let the inventor speak for himself :—

" The ' Cupron-Element ' is the outcome of years of research, commencing with the copper oxide element of Lalande and Chaperon, which was already in use in 1884. Several inventors, including Edison, have devoted their attention from time to time to improvements on the Lalande and Chaperon Element, partly from the point of view of a cheap and convenient source of electricity and partly as a ' reversible combination ' or an accumulator. These attempts have been mostly unsuccessful, especially with regard to the anode of the element, *i.e.* the copper oxide plate. This difficulty has however been overcome at last, mainly in consequence of a method of manufacture which produces a coherent and, at the same time, highly porous copper oxide plate, which, in its present state of perfection, can be entirely regenerated after exhaustion by simple exposure to the air; thus, the oxygen of the air furnishes the depolarization, which in most primary batteries is achieved by expensive and corrosive chemicals.

" The kathode of this element consists of zinc, which is absolutely unattacked by the electrolyte when the element is not in use. The electrolyte consists of commercial caustic soda. The plates are not removed from the solution until the cell is completely exhausted, and this, according to the work for which it is used, may represent periods of months, even a year; during the whole of this time the element is always ready to furnish a powerful and constant current for all purposes. The constancy of the current is only comparable to that furnished by an accumulator, which however is incapable of retaining its charge for long periods like the 'Cupron-Element.' The consumption of material is very low. About 1·25 to 1·5 grammes of zinc only are consumed per ampère hour furnished, and about 4 grammes of caustic soda. There is no depolarizer required,[1] and there is no attention necessary—not to say labour—which is so disagreeable a feature of most primary batteries. The 'Cupron-Element' has a somewhat low voltage (·85), but this is counterbalanced by its extremely low internal resistance."

The "Cupron-Element" is manufactured in the following types and sizes:—

Type.	Discharge Rate. Ampères.	Ampère hours.	Internal Resistance. Ohms.	Weight without Electrolyte.		Measurement.					
						Millimetres.			Inches.		
				Kilog.	lbs. ozs.	Height.	Width.	Length.	Height.	Width.	Length.
I.	1 to 2	50	·06	1·5	3 5	190	55	170	7½	2⅛	6¾
II.	2 to 4	100	·03	3·1	6 13	190	85	280	7½	3⅜	11
III.	4 to 8	200	·015	5·3	11 11	200	130	280	7⅞	5⅛	11
IV.	8 to 16	400	·0075	9·	19 14	250	140	370	9⅞	5½	14½

[1] Beyond that existant in the negative plate itself.

To give the reader an idea of the great constancy of this cell, when used either for a low rate or a high rate of discharge, we subjoin two tables, wherein the vertical lines show the time in hours, for which the current was drawn off; the horizontal lines giving the ampères and volts (in tenths), the curved lines showing the values and the variations in current and in E.M.F., during the entire trial, which extended over 400 consecutive hours at the low discharge rate, and over 35 hours at the high discharge rate :—

DISCHARGE CURVE OF TYPE I. (LISTED AT 50 AMPÈRE HOURS).

Electrolyte (Commercial Caustic Soda) ... ... 200 gr.
External Resistance, measured ... ... 5·34 ohms.
Internal Resistance, calculated ... ... 0·06 ,,
Average C.C. Voltage at Terminals ... ... 0·80 volt.
Average Rate of Discharge ... ... 0·15 amp.
  Efficiency 98·5 per cent.
                                Capacity 60 ampère hours.

TABLE I.

LOW RATE DISCHARGE.

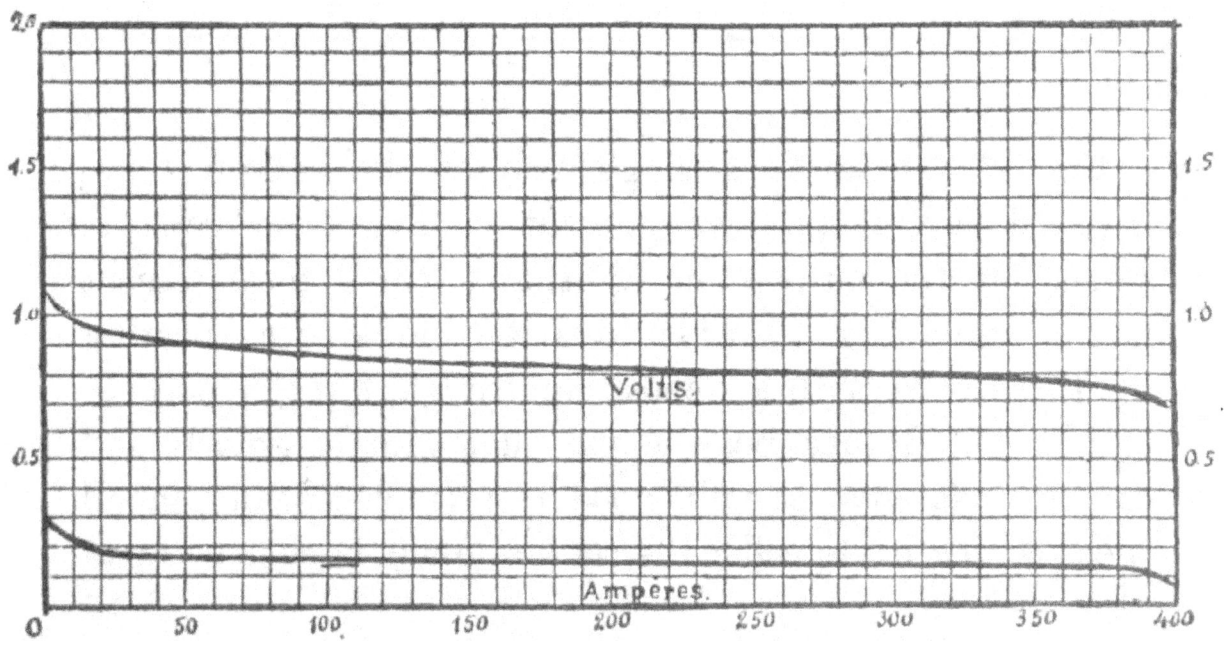

Electrolyte (Commercial Caustic Soda) ... ... 200 gr.
External Resistance, measured ... ... 0·43 ohm.
Internal Resistance, calculated ... ... 0·06 ,,
Average C.C. Voltage at Terminals ... ... 0·76 volt.
Average Rate of Discharge ... ... 1·55 amps.
Efficiency 86 per cent.
Capacity 53·5 ampère hours.

TABLE II.

HIGH RATE DISCHARGE.

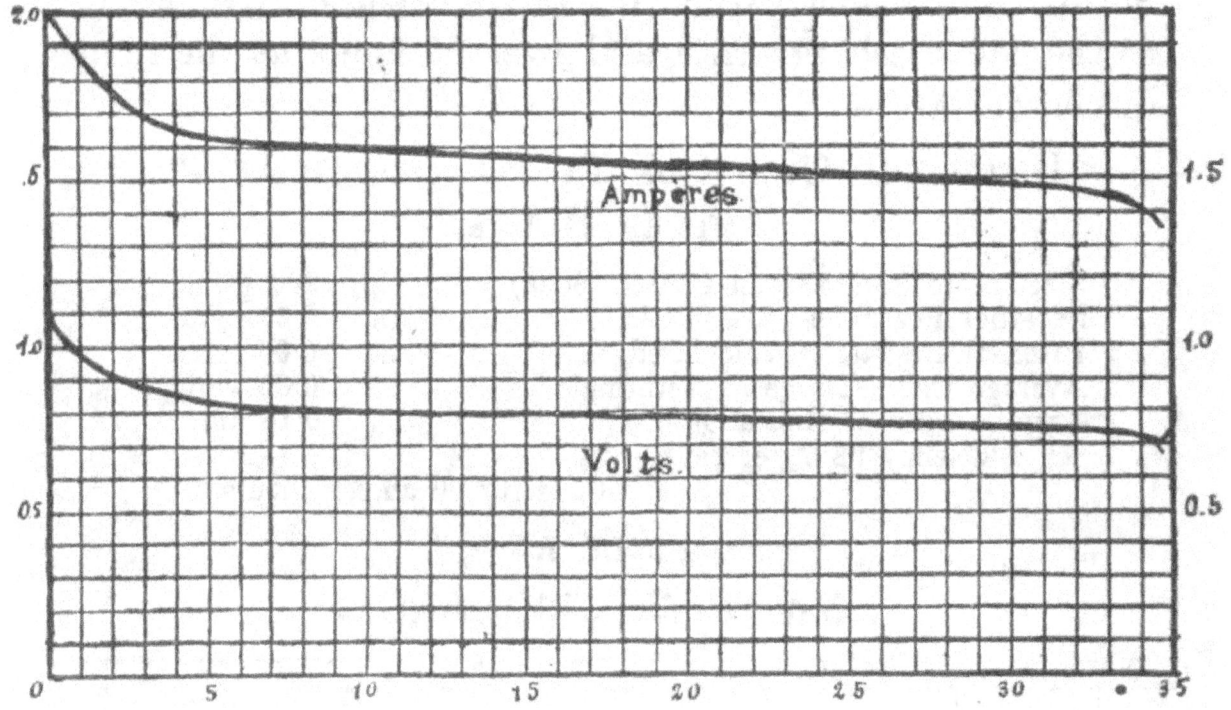

Fig. 75 illustrates a single cell of the cupron type; Fig. 76 shows the arrangement of five such cells in series, for the purpose of lighting a small incandescent lamp; while Fig. 77 illustrates the mode of connecting up three cells, with a deposition trough, for the purpose of plating. It has been found that caustic potash and caustic soda give equally good results electrically, in all these copper oxide cells: the sodium oxide is cheaper than the potassium oxide, but is rather apt to "creep"; but if the sides of the containing cell be paraffined, or be

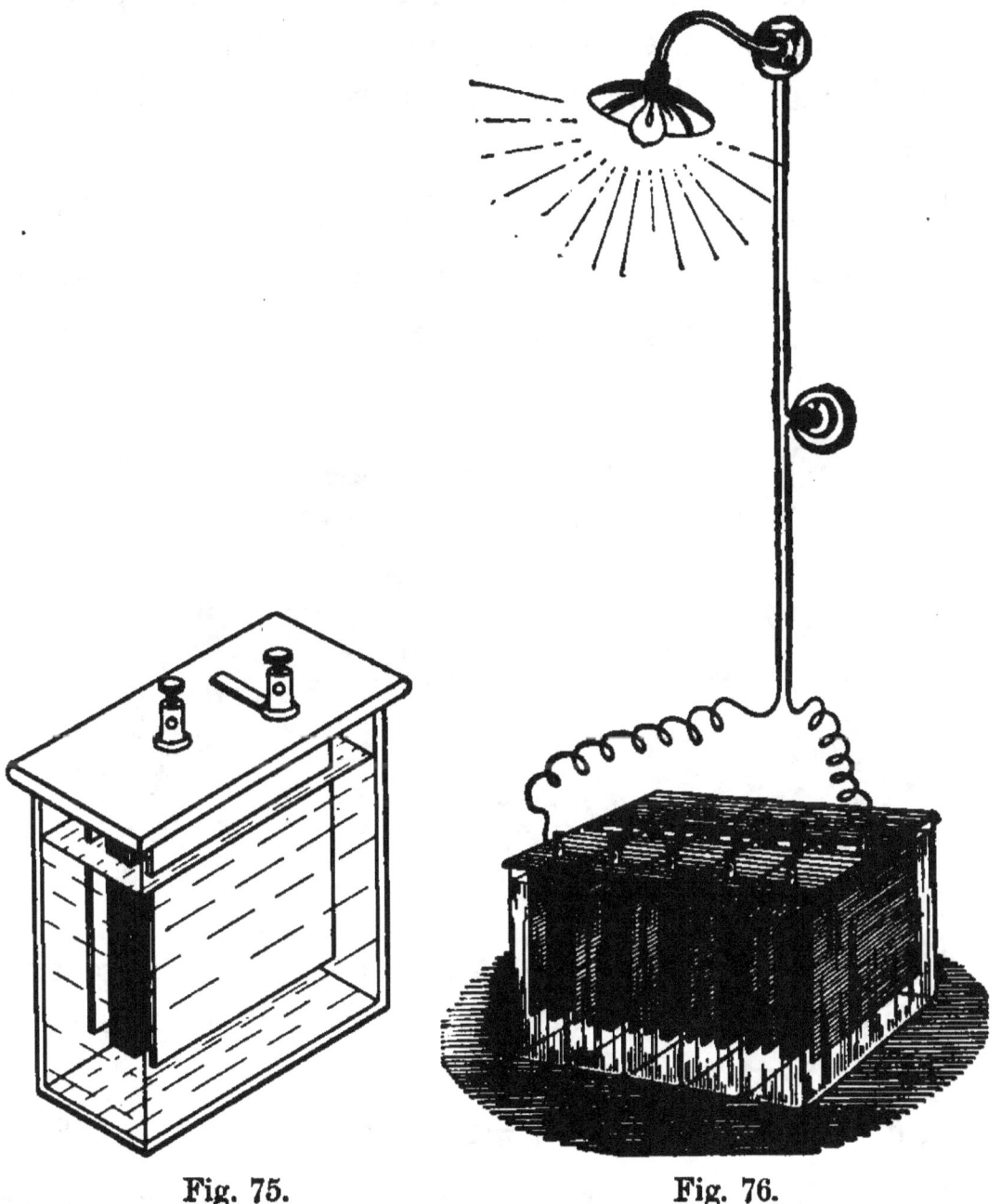

Fig. 75.   Fig. 76.

wiped over with heavy petroleum, even this slight disadvantage disappears. The copper oxide cells, as a class, may all be used as accumulators; or, in other words, by sending through them, when spent, a current from a generator, in the opposite direction, the original constituents are re-formed, and the cell is again capable of giving current. The following is the reaction which takes place during the discharge of a copper oxide, potash, and zinc cell :—

$$2\,Zn + 2\,KHO + 3\,CuO = 2\,ZnKO_2 + 3\,Cu + H_2O.$$

If soda be used instead of potash, the reaction becomes :—

$$2\,Zn + 2\,NaHO + 3\,CuO = 2\,ZnNaO_2 + 3\,Cu + H_2O.$$

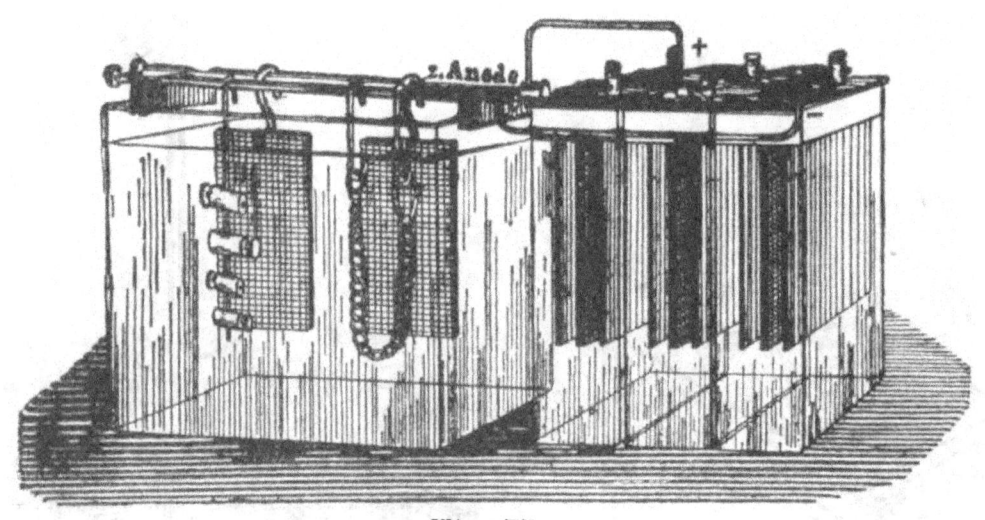

Fig. 77.

If current be sent through the cell in the reverse direction, this work is undone, and the following reaction takes place :—

$$3\,Cu + H_2O + 2\,ZnNaO_2 = 3\,CuO + 2\,NaHO + 2\,Zn.$$

§ 92a. *Bennett's*, or the " Tin-pot " cell (1882), is another form of alkaline excitant battery. As originally devised, it consisted in a Swiss milk-tin, 3″ high, 3″ in diameter, in the centre of which was placed a porous pot 4½″ high,

2″ in diameter. The interstitial space is packed round with iron borings. A zinc plate or rod is placed in the porous pot. Both compartments are filled up with a saturated solution of caustic potash. The E.M.F. is 1·14 volt; the internal resistance (in the above-mentioned size) 0·838 ohm. It is evident, that as the porous cell serves only to keep the iron borings away from contact with the zinc, that provided the iron borings be placed at the bottom of the cell, and the zinc be suspended some way above them, the porous cell may be omitted. In this case, a layer of heavy petroleum oil, floated on the surface of the caustic potash solution, will greatly increase the "life" of the cell, by preventing the carbonic acid gas of the atmosphere from converting the caustic potash into carbonate. This homely cell has been well spoken of for power of recuperation. It is closely allied to the Walker Wilkins cell, q. v.

§ 93. *Dr. Pabst* in 1883 described and patented in England a single-fluid cell, in which the elements are either carbon and tin, or carbon and iron. In the former case, the excitant and depolarizer is a 10 per cent. solution of stannic chloride in water; in the latter, the solution contains ferric chloride instead of the tin salt (Fig. 78). The E.M.F. of this cell is low; some authorities give it as 0·5 volts, others state it to be as high as 0·78 volt. In his claim, the patentee states that atmospheric oxygen really acts as the depolarizer, "as the aqueous solutions of the higher chlorides are reduced by the metallic electrode to the lower degree of oxidation" (*sic*), and again becomes oxidized to the higher degree, by the "absorption of oxygen from the atmosphere. The electrolytic liquids thus act continuously." As the ferric or stannic chlorides are acted on by light, chloride gas being evolved, it is advisable to blacken the containing jars, or otherwise protect the contents from the action of sunlight. When ferric chloride

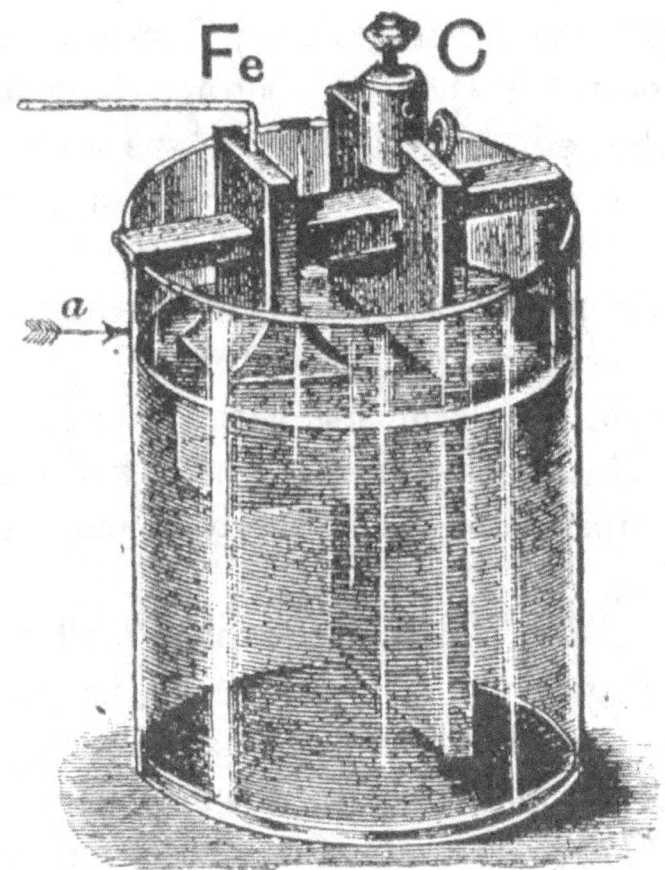

Fig. 78.

is used, the cell is continuously getting crusted up with ferric oxide, and this requires to be constantly removed. The action of the cell at first is: $Fe \div Fe_2 Cl_6 = 3 Fe Cl_2$; that is, one equivalent of iron in contact with one equivalent of ferric chloride gives rise to three equivalents of ferrous chloride. By exposure to the atmosphere, the ferrous chloride is partially oxidized, with the re-formation of ferric chloride, thus :—

$$O_3 + 6 \, Fe \, Cl_2 = Fe_2 \, O_3 + 2 \, Fe_2 \, Cl_6.$$

Dr. Conrad Pabst in 1887 patented in Germany another form of recuperative battery, which we have already noticed at § 84, as containing zinc oxide blocks, pitted against zinc, the excitant being zinc chloride. Both these cells seem to have fallen into disuse.

§ 94. *Mr. Anderson* (1889) brought out two cells, one a double fluid (described later on) and another a single fluid, which contains an organic chromium salt, to wit, the double oxalate of potassium and chromium, as the depolarizer. To prepare this salt he gives the following directions:—"Take a saturated solution of potassium bichromate, and add thereto oxalic acid, until effervescence ceases. The resulting solution is then to be slowly evaporated, until crystals of the double salt are formed; these are collected for use." Equal quantities of the saturated solution of this potassium chromium oxalate and of hydrochloric acid are mixed together; and in this mixture are introduced the zinc and carbon plates, forming the two elements of the cell. The action, owing to the strength of the solution employed, is very intense; therefore the battery is not fitted for long runs. We may note here, that hydrochloric acid may, in almost all cases, be substituted for sulphuric acid in battery work, with actual advantage as far as the production of a large current is concerned; but owing to the continual evolution of the fumes of hydrochloric acid gas, to the great detriment of all the brass-work about the battery, or in its proximity.

§ 95. *J. S. Wright*, in 1848, constructed a bichromate cell in which the negative element consisted in a long round rod of coke, on the upper portion of which was tightly bound a coil of copper wire, to form a connection; a length of this wire being left free. To protect that portion of the wire which was bound round the coke from the action of the exciting fluid, it was covered over with melted beeswax, which when cold was heavily varnished. The positive element took the form of a hollow cylinder of amalgamated zinc, in the interior of which the coke rod stood, kept from contact with the zinc by means of varnished cork-wedges.

§ 96. In *Ponci's* "circulating battery" (1879) zinc and carbon actuated by a solution consisting of 3 parts of potassium bichromate, 20 parts of water and 6 parts of hydrochloric acid, were employed. A series of rectangular leaden channels, beaked at one end, was placed in a slanting position, with the beak of the upper stood just over the wide end of the next below, and so on. In each channel was an amalgamated zinc plate, and above it, a carbon plate insulated from it by two rubber rings. The carbon was perforated at the point just under the beak or spout above it. The connections were made between the carbon of one channel or "cell," to the lead of the next, and so on for the whole series. The solution was supplied to the highest cell by means of a suitable syphon. Dropping thence from cell to cell, it was finally caught in a recipient after circulating through the entire series. If not entirely exhausted of the bichromate, the used solution could be returned to the upper containing vessel, so as to circulate once more.

§ 97. A battery very similar to this, in which, however, not only is the liquid caused to circulate, while the battery is in action, but the elements can be raised out of the

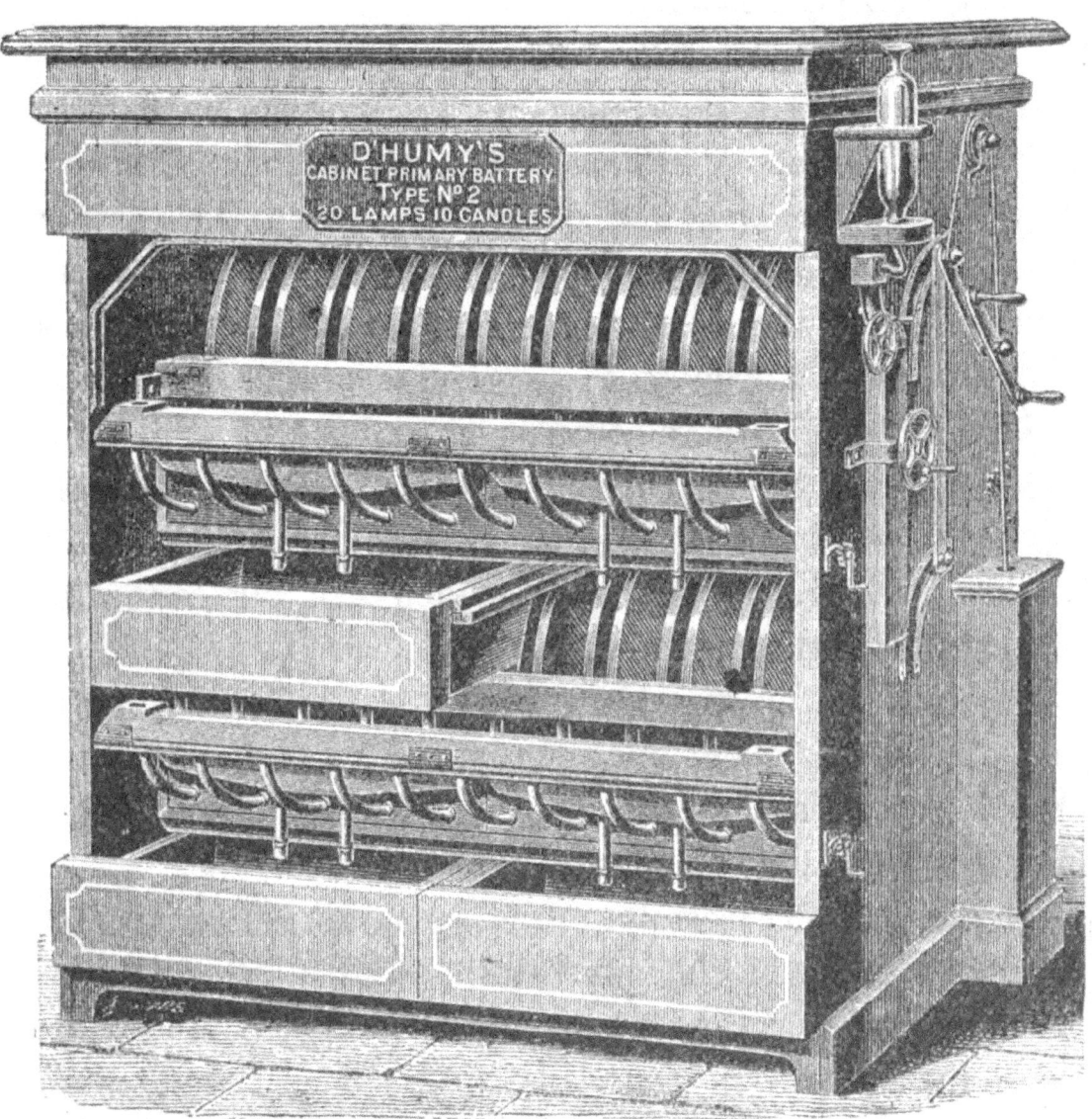

Fig. 79.

solutionby a quarter revolution of a handle, is *D'Humy's,* shown at Fig. 79. We give this as typical of a form of battery which is at once " plunge " and " circulatory." Here the elements (as shown) take the shape of quadrants of a circle, and are counterbalanced by a cord and weight, running in a box. Owing to their great cost, as compared with the dynamo, all these forms are, for lighting purposes at least, entirely obsolete.

§ 98. *Rowbotham Cell.* Even so lately as 1897 an attempt was made to supersede the dynamo by battery, and we give a short description accompanied by cuts to show the requirements of a battery-worked installation. The ROWBOTHAM battery as shown in Fig. 80 consists in a series of cells divided into two compartments. One of these compartments is open to the air and contains iron plates which form one of the electrodes of the battery; the other, a closed compartment, consists of two chambers connected together by rows of porous porcelain tubes in which are fitted carbon rods or pencils, which are all connected together, and thus form the other electrode.

The cells are arranged in rows on an inclined plane so that each cell is about one inch below the one above; the object of this is that the solution in each compartment may flow from cell to cell.

The solution or electrolyte used is contained in a suitable tank at the top of the generator, and is allowed to flow into the closed compartment of the top cell, and from there through an inverted ∪ connecting piece into the closed compartment of the next cell, and so on to the bottom cell and out into the exhaust or waste pipe.

Each set of three cells is provided with a supply of water from a tank connected to the cells by means of pipes; the water after passing through each set of three cells is allowed to run off into the waste. The action of the cell

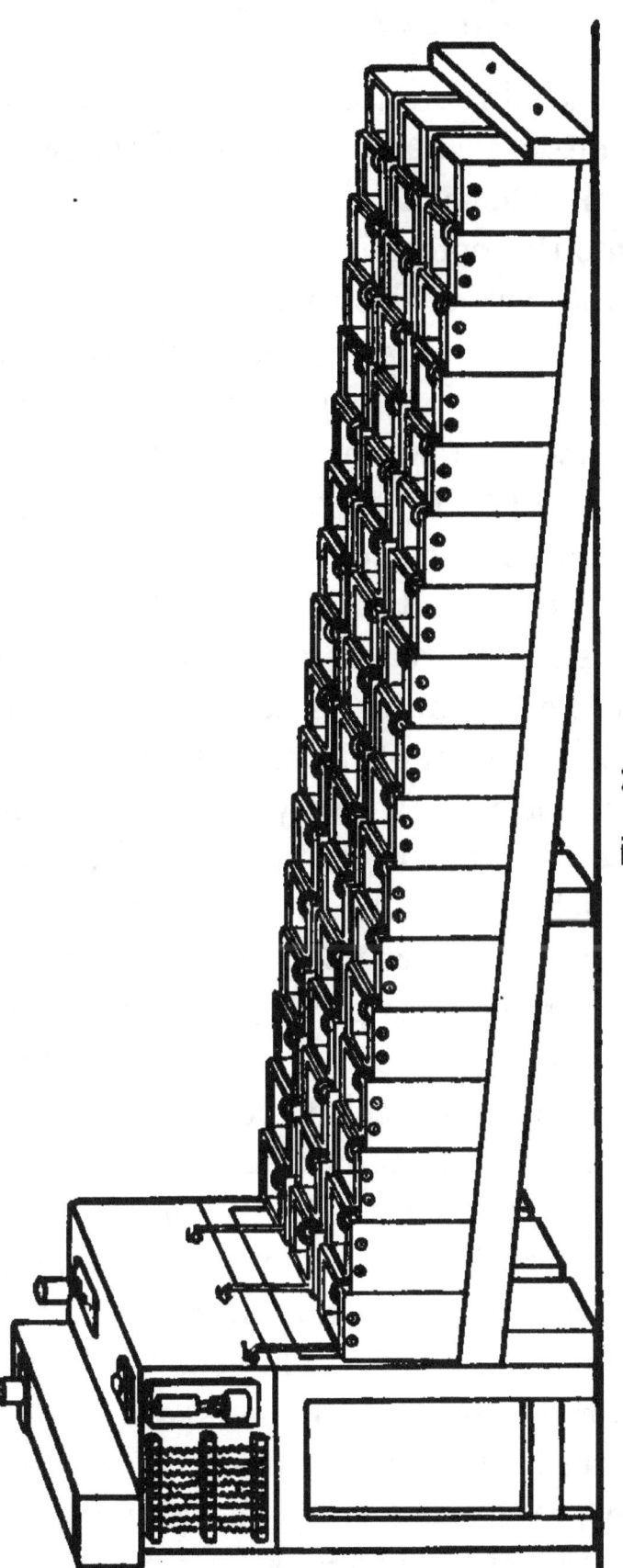

Fig. 80.

is as follows : a solution consisting of 30 parts sulphuric acid, 2 parts nitric acid, and 68 parts of water is placed in the upper tank, whence it flows into the closed compartments and into the porcelain tubes connecting them, thus surrounding the carbon electrodes. The open part of the cell having been previously filled with water, enough of the solution percolates through the porous cylinder as to slightly acidulate the water, thus setting up a potential difference of from 1·30 to 1·35 volt per each cell ; but such a small quantity of solution percolates through that very little consumption of the iron takes place. Nevertheless, owing to the large surface area of the electrodes, the internal resistance of the generator is sufficiently low to allow of 10 ampères being taken off from each cell at a pressure of about one volt per cell. Directly current is taken from the generator, gas is given off from the carbon electrodes of the cells, which gas being in the closed compartment causes a pressure to be set up, which tends to displace the solution, thus forcing it to percolate through the porcelain cylinder into the open part of the cell, so lowering the internal resistance of the cell to such an extent that up to 100 ampères may be taken off from each cell at a potential difference of one volt per cell. When the current is no longer taken from the generator, no more gas is evolved, the pressure therefore ceases, and no more solution is forced into the open compartment. The cells can then be flushed out with water, when they will be again in their normal condition of rest. The size of iron plates in each cell to give the above results is about 2 feet square.

§ 99. A circulating battery very similar to the above

is described in *The Scientific American* for September 1881. It consists in zinc plates 15 inches × 4¾ inches, wrapped in a flannel envelope moistened with sulphuric acid and placed between copper sheets 18 inches × 10½ inches which have been previously coated with calcined lamp-black made into a paste of sulphuric acid, and bent into a **U** shape. Each pair is then wrapped in a sheet of paraffined paper and placed in a trough of sufficient size to contain them. The elements are then connected as usual in series, and a stream of solution allowed to flow slowly through the trough. The solution consists of ¾ lb. of bichromate of potash, 1 lb. of sulphuric acid, and 1 gallon of water; 100 such elements will stand in a trough 3 feet × 1 foot × 2 feet, and give a very good light, equal to about 8 to 10 ampères at 50 volts pressure.

§ 100. *Coppinger's* "lifting" battery, for cautery, is another modification of the bichromate type, made with the special view of obtaining very large momentary currents, for cautery purposes. It was described in 1884, and consists as usual of zinc and carbon elements in a bichromate solution. A series of square zinc and carbon plates are cut so as to leave projecting lug at the top left-hand corner of the former, and the top left-hand corner of the latter. These two sets of plates are arranged at a little distance apart, on two separate stout brass rods, which pass through the respective lugs, so as to form virtually two large elements, one all carbon, the other all zinc. Washers are placed on the rods, to keep the plates apart. When not in use, the zinc plates are swung out of the solution by means of a heavy iron counterpoise

Except that there is no provision made for obtaining a circulation of the fluid (which is not needed for the short time required for cautery) this is very similar to the D'Humy's, previously described : see § 97.

§ 101. In 1888 *Rénard* devised a very light form of cell and battery specially driving the electro-motors of

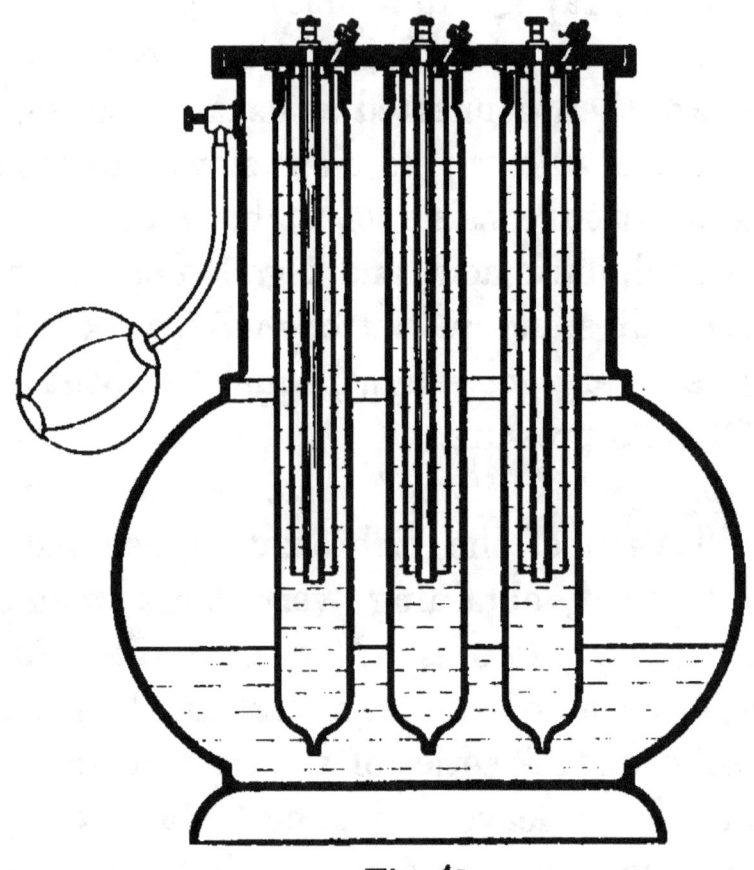

Fig. 81.

navigable balloons. The elements are, unamalgamated zinc, and platinized silver, in a mixture of sulphuric, hydrochloric, and chromic acids. As shown in Fig. 81, each cell consists of an ebonite or glass tube, of a length equal to ten times its diameter, for the sake of obtaining a large cooling surface. The tube is constricted at its lower extremity, so that a very narrow opening only remains at the bottom. A number of these cells are arranged and attached to the cover of an air-tight vessel, to which is fitted a small air compresser. The positive electrode (a thin platinized silver tube), and the negative electrode (a rod or wire of unamalgamated zinc), are supported from the cover, the zinc being inside the silver tube, but not touching it. The best excitant was found to be one containing 60 parts chromic acid to 200 parts of hydrochloric acid, specific gravity 11° Beaumé; which was found to give 1,600,000 joules for each pound of solution. Rénard's lightest battery consisted of 36 cells, ¾ inch in diameter, weighed charged 11 lbs., and gave half horse-power for twenty minutes.

§ 102. *Sicard* and *Fallé* devised a peculiar form of single-fluid cell, in which the "carbon" plate is made up of a paste consisting of powdered carbon, alum, sulphuric acid, and potassium bichromate compressed and dried. The positive element is as usual of zinc, and the two elements are immersed in a solution of ordinary salt 8 parts, water 100 parts. As the "carbon" plates were found to disintegrate, the mixture was finally packed and rammed into perforated carbon cylinders. In this form the E.M.F. was found to be 2·03 volts. This cell never came into extensive use. It is very similar in its behaviour to the "Fuller," *q. v.*

§ 103. There have been invented or proposed several other forms of single-fluid batteries, which have no practical interest: several of these will be treated of in a section set aside for "curiosities." We now pass to the consideration of "double-fluid" cells. By this term we understand all such cells as contain two fluids, one the excitant and the other the depolarizer, separated from each other by a porous septum. The most important of these is the "*Daniell*," which was described at § 49. We proceed to study the many modifications which the Daniell cell has undergone at the hands of different experimenters, in their endeavours to fit it for certain special purposes. In the original DANIELL CELL the constituents were, zinc in dilute sulphuric acid, and copper in a saturated solution of copper sulphate. The zinc was a cylinder standing on a porous cell filled with diluted sulphuric acid by a small funnel through a wooden top. The porous cell had also a drain-pipe from the bottom, which was brought up outside the cell so as to show the level of the solution inside. The porous compartment itself stood on an outer cell made of copper, which acted as the positive pole of the cell. The solution in this consisted of water having as much copper sulphate in solution as it could take up. The porous cell may be made of unglazed porous porcelain, or of lighter material such as parchment, or even of brown paper. The zinc cylinder was amalgamated. There was a cage containing crystals of sulphate of copper in the upper part of the outer cell to supply the solution as it weakens.

When working, the action is as follows:—The sulphuric acid acts on the zinc, forming zinc sulphate, which dissolves, and hydrogen, which endeavours to reach the positive copper, but on its way it has to pass through the strong solution of copper sulphate. A re-action occurs, resulting in the displacement of the copper in the sulphate by the hydrogen, and the copper is deposited on the copper cylinder, which therefore increases in weight by the working of the cell. Thus the sulphuric acid in the zinc compartment is replaced by zinc sulphate, which, however, works nearly as well. The copper sulphate solution becomes charged with sulphuric acid, and hence requires replacing periodically. The cell then consumes zinc and copper sulphate. E.M.F. 1·079 volt. Internal resistance varies, but usually is rather high. The cell gives a very constant current of low intensity, and was much used for telegraphy. It was first used by Daniell in 1837, called by him the "Constant" cell. There is a modification having the copper in the inner cell and the zinc in the outer one. The copper rod carries a little sieve in which are placed the copper sulphate crystals.

§ 104. *Kramer Cell.* Constituents :—Same as Daniell's cell. Arrangement :—The copper electrode is divided into two cylinders. One of these is placed in a diaphragm which is filled with copper sulphate solution, and which is put inside another diaphragm containing dilute sulphuric acid. The zinc cylinder is placed outside the second diaphragm. Provided the cell has sufficient copper sulphate in it, it will work for months without attention. The object of two coppers is to prevent deposition of copper in the diaphragm.

§ 105. *Trough Daniell.* Arrangement :—A rectangular box divided into cells by slate partitions. Each of these cells is again divided into two parts by a porous porcelain plate. One of these divisions has a solution of copper sulphate and a copper plate, and the other has either water or a very dilute solution of zinc sulphate and a zinc plate. The zinc plate and the copper plate are joined by a copper strip which is bent so as to allow both plates to hang from the slate partitions. With a sufficient supply of copper sulphate crystals the battery will work for a month without attention. The zincs are usually *cast*, which are not so good as rolled. It is also difficult to make a good joint with slate and wood. This is also called the Daniell Trough Battery, and is much used in the English Telegraph Service. The trough is usually of teak, and all joints marine-glued to render them watertight.

§ 106. *MacDonald Cell.* Constituents :—Are the same as Daniell's cell. Arrangement :—An outer cylindrical vessel of earthenware, 4 inches in diameter, 12 inches high. Directly inside this is a cylinder of copper gauze ; or even a spiral of No. 16 copper wire, reaching to the top of the cell, and forming one electrode. Inside this is placed a porous pot containing a zinc rod. The space between the outer and inner vessels is packed with coarsely-powdered coke. The excitant in the porous cell is a solution of 1 lb. of common salt in one gallon of water. The liquid in the copper-coke compartment is a solution of 1 lb. of copper sulphate in one gallon of water. Resistance of a single cell, this size, said to be ·14 ohm, but is really much higher ; E.M.F. one volt. It is virtually a poor Daniell cell.

§ 107. *Fleming's Standard Daniell Cell.* Constituents : —Same as Daniell's. Arrangement :—A ∪ tube, ¾ of an inch in diameter and 8 inches long, has a tubulure with a glass tap proceeding downward from the bend.

The upper ends of both limbs have rubber stoppers, through which the electrodes project. Branching from each limb, just below the stoppers, are tubes which lead by a glass top into two bulbs open at the top. There is also a tube with a stop-cock in the lower part of one limb, leading to a vessel placed below. The whole is mounted on a vertical stand. To set up the cell, starting with it empty, and both taps leading to the top bulbs shut, the left-hand bulb is filled with zinc sulphate solution, and the other with copper sulphate solution. The tap of the zinc solution is now opened, and this allows the U tube to be filled with the zinc sulphate solution; the tap is then closed.

The zinc rod, usually kept in a small tube mounted on the stand, is put in place in the left-hand limb, and tightly corked up. Then the tap in the lower part of the opposite limb is opened, which lowers the level of the solution in that limb only. The tap of the top right-hand bulb is opened to admit the copper sulphate solution, which takes its place above the zinc sulphate in that limb. The copper rod is then put in. The sharp line of separation of the two may be established by drawing off the mixed layer. Further particulars:—Oxidation of the zinc lowers the E.M.F., oxidation of the copper raises it. With solutions of equal specific gravity the E.M.F. is 1·104 volt. If the specific gravity of the copper sulphate solution is 1·1, and that of the zinc sulphate 1·4, both at 15° C. (59° F.) the E.M.F. will be 1·074 volt. The copper is prepared for use by freshly electro-plating it with copper; the zinc is clean and pure.

§ 108. *Dr. Sloane's Standard Daniell Cell.* Two test-tubes employed for the solutions, with a syphon to connect them.

§ 109. *Lodge's Standard Daniell Cell.* Arrangement:—A wide-mouthed bottle with a wood or other stopper, has a glass tube open at the top and bottom put through it. In this is placed a pure zinc rod. To the bottom of the tube a small glass tube containing copper sulphate crystals is secured by a rubber band or by string. The bare end of a gutta-percha-covered copper wire projects from the bottom through a glass tube in a well-fitting cork, forming the copper plate. The cell is partly filled with zinc sulphate solution. Internal resistance is so high that it is only used in zero methods with a condenser.

§ 110. *Sir W. Thomson's Standard Daniell Cell.* Arrangement:—A zinc disc is placed at the bottom of a cylindrical vessel and a solution of zinc sulphate of specific gravity 1·2 is poured over it. By means of a funnel and bent tube a half-saturated solution of copper sulphate is poured over this, and floats on it. E.M.F. is 1·072 *true* volts at 15° C.

§ 111. *Sir W. Thomson's Tray Cell.* Arrangement:—A copper plate is put at the bottom, and a saturated solution of zinc sulphate is poured over it. The zinc is in the form of a grating, and is placed horizontally near the surface of the solution. A glass tube is placed vertically in the solution, with its lower end just above the surface of the copper plate. Crystals of copper sulphate are dropped down this tube, and, dissolving in the liquid, form a solution of greater density than the zinc sulphate solution alone, so that the copper sulphate can only reach the zinc by diffusion. To retard this process of diffusion, a syphon consisting of a glass tube stuffed with a cotton wick, is placed with one extremity midway between the zinc and copper, and the other in a vessel outside the cell, so that the liquid is very slowly drawn off near the middle of its depth. To supply its place, water or a very dilute solution of zinc sulphate is added above when required. In this way the greater part of the copper sulphate rising by diffusion is drawn off before it reaches the zinc, which is thus surrounded with a solution nearly free from copper, and having a slow downward movement, which farther tends to prevent the copper from rising. During action, copper is deposited on the copper plate, and the freed sulphion radical (SO_4) rises to the zinc, where it forms zinc sulphate. Thus the upper solution gets heavier, and the lower lighter. To prevent this from disturbing the strata, the tube must be kept well supplied with copper sulphate crystals.

§ 112. *Siemens' and Halske's Cell.* Constituents: —Same as in a Daniell's cell. Arrangement:—At the bottom of a cylindrical vessel, a copper spiral or plate is placed, and from the centre of this a copper wire leads upwards for external connection. Over the plate is a porous clay vessel, having a glass tube fastened to it. The space over the clay vessel is filled up to about half-way just over the shoulder of the clay vessel, with a prepared paste consisting of paper pulp and sulphuric acid. The glass tube and clay vessel are filled with crystals of copper sulphate. Over the paper mass a piece of linen is laid, and a ring of zinc is placed on it, encircling the central glass tube. A brass rod and terminal make connection between the zinc and the outer circuit. The cell is charged by pouring acidulated water into both outer and inner vessels. Further particulars :—Internal resistance high, because of the paper mass. The current remains constant if the fluid surrounding the zinc is renewed every fortnight. The piece of linen should also be cleaned at similar periods. After use for some time, copper is found to have separated out at the diaphragm and in the pulp, and the copper sulphate sometimes reaches the zinc. The high internal resistance does not interfere with its use for telegraphic work.

§ 113. *Trouvé's Blotting-pad Cell.* Constituents :— Same as those in Daniell's Cell. Arrangement :—In a cylindrical vessel a copper plate is put at the bottom, covered with a pile of blotting-paper soaked with a solution of copper sulphate, on which is another set of blotting-paper soaked with a solution of zinc sulphate, and a zinc plate on the top of all. To set up the cell, water is poured on to the discs of paper (which have previously been saturated with their respective solutions and dried), until drops appear at the edges, when they are pressed together. The zinc has a wire which connects to the outer circuit, and the similar wire for the copper is covered with gutta-percha and passes through the blotting-paper and zinc, and is fixed to the lid. The copper sulphate is gradually used up, while zinc sulphate is produced. Useless consumption when not in use can be prevented by exposure to a draught after use.   This form of cell is especially recommended for military and medical use, and was invented to prevent useless consumption of zinc when not in action.

§ 114. *Minotto's Cell.*   Constituents:—Same as a Daniell's cell, but with a layer of sand instead of a porous cell.

At the bottom of an earthenware vessel is placed a layer of coarsely-powdered copper sulphate, and on this is a copper plate provided with an insulated copper wire. On this there is a layer of sand or sawdust, and then a zinc cylinder; water is then poured on the sawdust.   If sand be used it should be non-absorbent, such as quartz. Sometimes a zinc plate is laid on the sand direct, instead of a cylinder.   The wire connecting the zinc to the outer circuit should be coiled into a helix in order to allow the zinc to sink down as the crystals are consumed.   The resistance is higher than that of a Daniell, but the current remains constant for months.   Compared with a Daniell it uses less zinc and copper sulphate, and requires less attention, the occasional cleaning of the zinc and replacing of the evaporated water being excepted.   The connection-wire from the copper is gutta-percha covered.

§ 115. *Barley's Gravity Cell.*   Same as Minotto's cell, but the solutions are separated by their own specific gravities only; this cell has been very little used. Patented in 1885.

§ 116. *Varley's Cell* (Gravity).   Same as Siemens and Halske's, but having zinc oxide substituted for the paper pulp of the other cell.   Of no practical use.

§ 117. *Meidinger's Cell* (Gravity type).   Constituents:

—Zinc and copper or lead in water, or a solution of magnesium sulphate, and copper sulphate. Arrangement: —An outer vessel having a sudden contraction about two-thirds of the way down. On this rests a cylinder of zinc, which has a connecting-wire attached. A second smaller vessel is placed inside, and a sheet of copper or of lead is placed within the second vessel, the connecting-wire being well insulated. A tube having a small hole at its lower end, and whose top passes through the lid, dips into the second vessel with the copper or lead, and is filled with crystals of copper sulphate. To charge the cell the outer vessel is filled with water or a solution of magnesium sulphate (Epsom salts), to diminish resistance. The crystals of copper sulphate dissolve, and thus a solution of copper sulphate surrounds the copper or lead electrode. Copper sulphate having a higher specific gravity, falls to the bottom, and raises the lighter salts solution to the upper part around the zinc cylinder. When employed for telegraphy, as it is in Germany, the copper sulphate requires frequent supplies, for which reason Meidinger devised his :—

§ 118. *Balloon or Flask Cell.* Constituents :—Same as the original form of Meidinger's cell. Arrangement: —The tube containing the copper sulphate crystals is replaced by a flask, which is filled with crystals of copper sulphate and water, usually about 2 lbs., and inverted, so as to remain full and act as a cover for the vessel. The E.M.F. is about the same as that of a Daniell, and its resistance is much lessened by omitting the diaphragm.

§ 119. *Post-Office, Standard Daniell Cell.* Constituents:—Zinc in zinc sulphate, and copper in copper sulphate, solutions. Arrangement:—A box having three distinct receptacles, the left-hand one containing a plate of zinc immersed in water; in the right-hand one is a rectangular flat porous pot containing a plate of copper in saturated copper sulphate solution. The porous pot is immersed in water. These two are only receptacles for the zinc and copper when not at work. The centre one is nearly filled with a half-saturated solution of zinc sulphate, and has a little zinc cylinder fitted in a special compartment at the bottom. When the cell is to be used, the porous pot is put in the central compartment, in which the zinc is also placed. Both the zinc and porous cell are replaced in their resting-compartments when the cell is no longer required in use. The small quantity of copper sulphate which passes through the porous cell during action is reduced by the little zinc cylinder, and so keeps the zinc sulphate solution pure. In good conditions a fresh cell gives 1·079 volt, but in ordinary working it is reckoned at 1·07 volt.

§ 120. *Erhard's Circulating Battery.* Constituents: —Zinc in water and lead in copper sulphate solution. Arrangement:—Many cells are grouped together like a trough battery, and each cell is formed by the plates of adjoining cells as sides, and a three-sided papier-maché frame as ends and bottom. The electrodes are a rectangular plate of zinc covered with lead-foil on one side and with cloth as a diaphragm on the other side. The battery is built up of these compound plates by placing a papier-maché frame between each pair, and tightening the whole up between two boards and bolts passing through the whole. There is a glass vessel over the battery, containing crystals of copper sulphate and some water. Two tubes pass out through a tubulure at the bottom, each leading to a distributing pipe with spouts opening into each cell. The tubes pass out of the vessel one above the other. The battery, when to be set up, is filled with water. The two taps leading to the overhead vessel are opened, and the heavy copper sulphate solution flows down and displaces the water, which flows into the other vessel by the second tube, and there dissolves a further quantity of sulphate. The lead plates are at once covered with a deposit of copper, and act as a copper plate. The battery is then practically a Daniell. As the zinc is dissolved the specific gravity of the ascending liquid becomes gradually higher, and finally circulation ceases, and the battery must be recharged. The largest size consists of 17 cells, and has a useful surface of 450·5 sq. cms. (69·75 sq. ins. approx.). One of these was subjected to a discharge of from 3 to 8 ampères ten times, and a total discharge of 305 ampère hours. The E.M.F. was, on the average, 15 to 16 volts. The internal resistance was less than 1 ohm for the greater part of each run, but rose at the end of each to 1·5 ohm. In the total of 70 hours' work 8·8 kilogrammes (19·36 lbs., 1 kilogramme = 2·2 lbs. approx.) of zinc, and 40 kilogrammes (88 lbs.) of copper sulphate, were used. The theoretical quantities were 6·3 kilogrammes of zinc and 24 kilogrammes of copper sulphate. Results of smaller cells were not quite so good.

§ 121. *Silver's Cell.* Constituents :—Same as ordinary Daniell. Arrangement :—A flat copper cell, having a piece of leather between two sheets of copper, on one side, perforated. Opposite perforated side the zinc plate is placed in water. The copper cell contains copper sulphate crystals, and is varnished inside.

§ 122. *Fuller's Trough Daniell.* Twelve compartments, each having a partition at half the total height. The zincs reach half the depth of cell, and the coppers the total depth (see also § 105).

§ 123. *Cruikshank's Trough Daniell.* Gutta-percha troughs, filled with sand and acid $\frac{1}{12}$ to water 1. Copper plate with copper sulphate below the sand, zinc and zinc sulphate above the sand.

§ 124. *Rowland's Magnesium Daniell Cell.* Constituents :—A plate or bar of magnesium coated with paraffin wax except on the lower end, and immersed in a solution of the sulphate (or other salt) of magnesium. Bar or plate of copper, in porous cell, with copper sulphate solution. Arrangement :—The magnesium is in a porous tube made of " parkesine," and is fed slowly down as it dissolves. The copper is best round outside. E.M.F. higher than with zinc.

§ 125. *Mullin's " Sustaining " Cell.* A central copper cylinder, with wooden bottom, containing the copper sulphate solution and crystals. A zinc cylinder is placed outside, in a solution of salammoniac or sulphuric acid, and separated from the copper by a membrane.

§ 126. *Himmer's " Variable " Cell.* Arrangement :— Outer vessel of glass, containing a small conical vessel in which the copper is put. The zinc is in the form of a ring sprung into the upper part of the outer cell. Over the cell is an inverted vessel full of water and copper sulphate crystals, and a tube, capable of sliding up and down in a cork closing the vessel, leads into the copper compartment of the cell. By altering the depth of the lower end of the tube, more or less of the copper solution can be allowed to flow in.

§ 127. *Kohlfürst's Cell.* Constituents :—Zinc and lead in zinc sulphate or magnesium sulphate and copper sulphate.   Arrangement :—A glass vessel having a constriction near the bottom, has a plate of lead nearly at the bottom, lying in crystals of copper sulphate, which are filled in up to the constriction. On this a perforated earthenware plate is placed. There is an iron lid to which is fixed a mass of zinc having a terminal attached. Also a funnel passes through the lid, reaching nearly to the earthenware plate, for the purpose of pouring in the solution of zinc or magnesium sulphate. Much used on the Buschtĕhrader railways, where it is said to have worked very economically.

§ 128. *Callaud's Cell.* Constituents :—Same as Meidinger's, of which it is a simplified form. Arrangement :—Glass or earthenware vessel, having a plate of copper at the bottom, and a cylinder of zinc hanging in the upper part by hooks on the rim of the vessel. This zinc only reaches down to about midway in the cell. On the plate of copper, which is connected by a gutta-percha insulated wire to the external circuit, is a layer of crystals of copper sulphate. The whole is filled with water, and the cell works constantly for some time if not agitated. Its action is very economical in copper sulphate, and it is easily manipulated. To set up the cell it has been found convenient to pour the zinc sulphate solution into the vessel first, until it nearly reaches the zinc cylinder, and then by means of a syphon to allow the copper sulphate to replace the zinc sulphate. Further particulars:—This cell is much used in France and the United States. Its usual dimensions are:—Outer glass, 8 inches high, 5 inches in diameter. The copper sheet has a surface of 15·5 square inches, and a height of 1·25 inches. The zinc sheet is $2\frac{3}{4}$ inches wide, bent to a circle of 2 inches radius. Every three months the zinc is cleaned, at the same time about 10 ounces of copper sulphate being added. To prevent the increase of resistance a portion of the zinc sulphate is removed and replaced by water. From independent observations by different persons during several years it has been found that a cell costs about $6\frac{1}{2}d.$ per year. Resistance varies from about 30 ohms to about 9 ohms.

§ 129. *Lockwood's Cell.* Constituents:—Same as Meidinger's and Callaud's. Arrangement:—A glass vessel, 15 inches high and $7\frac{1}{2}$ inches in diameter, is half filled with crystals of copper sulphate. The copper electrode consists of two spirals, one at the bottom and one at the top of the crystals, connected by a wire to the external circuit. The liquid used is a solution of zinc sulphate, and evaporation is lessened by a layer of oil. The zinc electrode is a massive circle divided into quadrants, and supported from the rim of the vessel just beneath the liquid.

§ 130. *Eisenlohr's Cell.*   A Daniell having a solution of sodium bitartrate instead of sulphuric acid.   Invented in 1849.

§ 131. *Reynier's Cell*, No. 1.   Constituents:—Zinc in sodium hydroxide solution, and copper in copper sulphate solution.   Arrangement:—The copper electrode is a flat sheet bent so as to fit down two opposite sides of the cell and across the bottom.   It has a slot in it.   The zinc is a sheet, inside a parchment cell containing a solution of sodium hydrate (caustic soda).   The solution outside, with the copper, is one of copper sulphate.   By the use of sodium hydrate instead of sulphuric acid, the diffusion of copper sulphate is prevented, and the E.M.F. is raised to from 1·3 to 1·5 volt, besides the consumption of zinc an open circuit being prevented.   Invented in 1880.   The caustic soda solution contains 300 parts of sodium hydroxide dissolved in 1000 parts of water; the copper sulphate solution has also some sulphate of sodium or sulphuric acid dissolved in it.   Both contain small quantities of certain alkaline salts, such as calcium chlorate, sodium chloride, ammonium salts, etc., to diminish the internal resistance.   The resistance of a 3-litre or 4·8 pints (1 litre = to 1·6 pints nearly) cell, with a parchment paper porous cell, is about ·075 ohm.

§ 132. *Reynier's Cell*, No. 2.   Constituents:—Zinc in zinc sulphate solution, and copper in copper sulphate solution.   Arrangement:—The copper electrode forms an outer containing-cell or trough of flat form, 9 inches high, 2 inches wide, and 16 inches long.   The zinc sheet is 4 inches shorter than the copper one, and is placed so as to leave a greater space at one end than at the other. At the short end is placed a basket of copper sulphate crystals, to supply as the solution requires it. The bottom of the cell has a wooden flooring. The zinc is surrounded by a wrapping of parchment paper. The zinc sulphate passes through the parchment by itself by the process known as "osmose." Only the copper sulphate requires renewal, and there is no sulphuric acid added, but the copper sulphate percolates through to the zinc; the re-action following results in formation of zinc sulphate which diffuses through into the copper compartment again. Every 24 hours one pint of solution is replaced by water, by a tube at the end of the cell. Invented in 1881. Further particulars:—In April 1882, E. Reynier had 500 such cells, 68 of which he used for charging secondary cells for lighting purposes. The work done by these 68 cells in eight hours was equivalent to one horse-power, at an outlay of thirteen shillings, or about thirteen times that produced by a steam-engine and dynamo.

§ 133. *Lead-Zinc Gravity Cell.* Constituents :—Lead and zinc in copper sulphate and water. Arrangement :— A heavy perforated cone of lead; prolonged upwards as a tube which passes through a wooden lid. This lead cone and tube, which form an inverted funnel, rests on the bottom of the vessel, and is filled with crystals of copper sulphate. A heavy ring of zinc is suspended from the lid, only slightly below the surface of the liquid. Water is poured in, and dissolves the copper sulphate, forming a strong solution at the bottom, leaving the upper part almost without copper. E.M.F. 0·8 volt; used for electric signs, bells, and models.

§ 134. *Perreur-Lloyd Generator.* An attempt at producing a saleable by-product to cheapen the current. Constituents:—Zinc in dilute sulphuric acid, copper in copper sulphate solution. Arrangement:—Copper plates in porous cells with copper sulphate solution; zinc outside

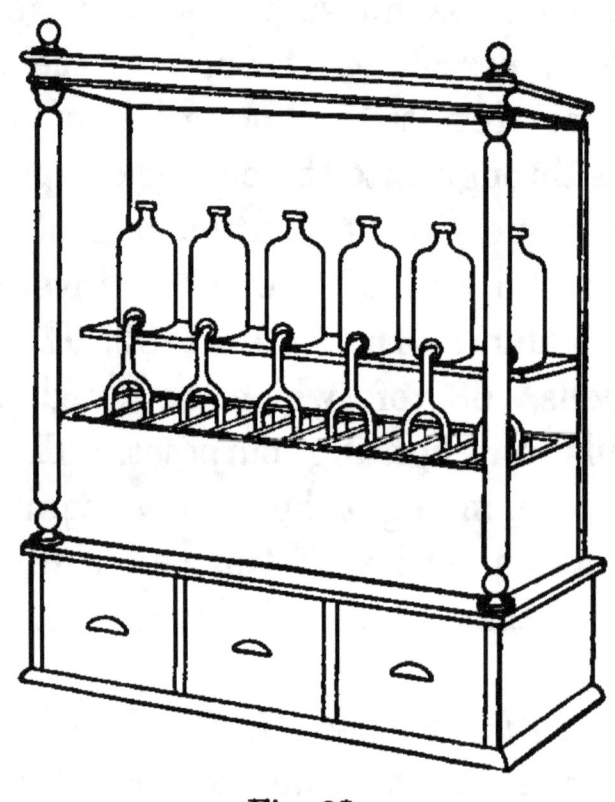

Fig. 82.

with the acidulated water. The zincs are calculated to last two months, and the solution of copper sulphate is kept in jars above the cells. The battery was sold, arranged on a handsome stand with reservoirs to collect the residual zinc sulphate in drawers; then next above were the cells, each copper cell fed by a separate glass reservoir; lastly, on the top of the stand, a set of accumulators, which absorbed the excess power of the battery at first, giving it out again when the battery became weak. Invented by Perreur-Lloyd et Fils, in 1891. See Fig. 82.

§ 135. We have already noticed, under the heading of "Agglomerate," that manganese dioxide constitutes a most useful depolarizer. We now proceed to describe the many forms of double-fluid cells, in which manganese dioxide is contained in a porous cell along with the carbon, while the zinc stands in a solution of sal-ammoniac, or similar salt. All these cells are modifications of the one known as the "*Leclanché*," from the name of the original patentee.

§ 136. *The Leclanché Cell.* Constituents:—Zinc in a solution of ammonium chloride, and carbon in manganese dioxide. Arrangement:—A zinc rod, amalgamated, in the outer jar, which is about half filled with a solution of ammonium chloride (sal ammoniac), whose strength should be about half saturate, as if it is saturated it "creeps" up the sides and causes electrical leakage. A carbon rod or plate is packed tightly into the porous pot, with a mixture of equal volumes of crushed carbon and pebble manganese dioxide. The top of the porous pot is covered with pitch, the leaden cap of the carbon plate, with its brass terminal projecting, and a small glass tube put through the pitch, to allow the escape of gases from the cell during action (see Fig. 83). This cell very easily polarizes, but quickly recovers itself. Its resistance varies with the concentration of the solution. With a cell exposing 8 square inches of negative surface, and 1 pint of solution of ordinary working strength the resistance is 1·5 ohms. The E.M.F.

is about 1·48 volt when not polarized, or 1·32 if the sal-ammoniac solution is saturated. The resistance may vary from 1 ohm when freshly put up to 50 ohms when nearly spent. The zinc should be amalgamated, rolled or drawn metal. If the cell be required for use immediately after making up, some of the solution should be poured into the porous pot, instead of waiting for it to percolate through. Leclanché patented this form of cell in 1868.

Fig. 83.

The reaction which takes place in the Leclanché cell is variously given. The usual view is :—

$$Zn + 4\,NH_4\,Cl, + 2\,MnO_2 = Zn\,Cl_2 + 2\,NH_4$$
$$Cl + NH_3 + Mn_2\,O_3 + H_2O.$$

Others view the action as taking place in three successive steps, viz. :—

$$Zn + 2\,(NH_4\,Cl) = Zn\,Cl_2 + NH_3 + H_2.$$

The zinc chloride and the ammonia are then supposed to react as follows :—

$$Zn\,Cl_2 + 2\,(NH_3) = 2\,(NH_2)\,ZnCl_2 + H_2.$$

Finally the free hydrogen seizes upon a portion of the oxygen of the manganese dioxide, to produce therewith water, thus :—

$$2H + 2\,(Mn\,O_2) = H_2O + Mn_2\,O_3.$$

If the sal ammoniac be not in sufficient quantity, the water reacts on the zinc, the solution gets milky, and the zinc becomes encrusted with a mixture of zinc oxide and oxchloride of zinc and ammonium. The addition of a little more sal ammoniac, after carefully scrubbing the zincs, will remedy this defect. The proper strength for the solution is about 4 ozs. to the pint of water.

§ 137. *Siemens' and Halske's Improved Leclanché* (1900). Arrangement :—In a cylindrical or rectangular cardboard receptacle, is placed a zinc vessel of similar shape, whose bottom is stopped either with zinc, or asphalte, firmly united to the zinc sides, so as to prevent any escape of liquid. Standing upright in the zinc vessel, is placed the carbon electrode, which is surrounded by a porous depolarizing mass secured to it by a gauze bag. The electrolyte, which is a concentrated solution of sal ammoniac, is contained in the space between the zinc and the carbon, and is sealed over with a layer of asphalte, so that the gas can only escape to the atmosphere by passing through the depolarizer, on which the passage of the gas exerts a continual loosening action, and thence through a gas chamber in the upper part of the cell, which is filled with husks of rice. Above this is a layer of asphalte, which closes the cell, and retains the parts in position. Small capillary glass tubes passing down through the asphalte, allow of the escape of gas, but not

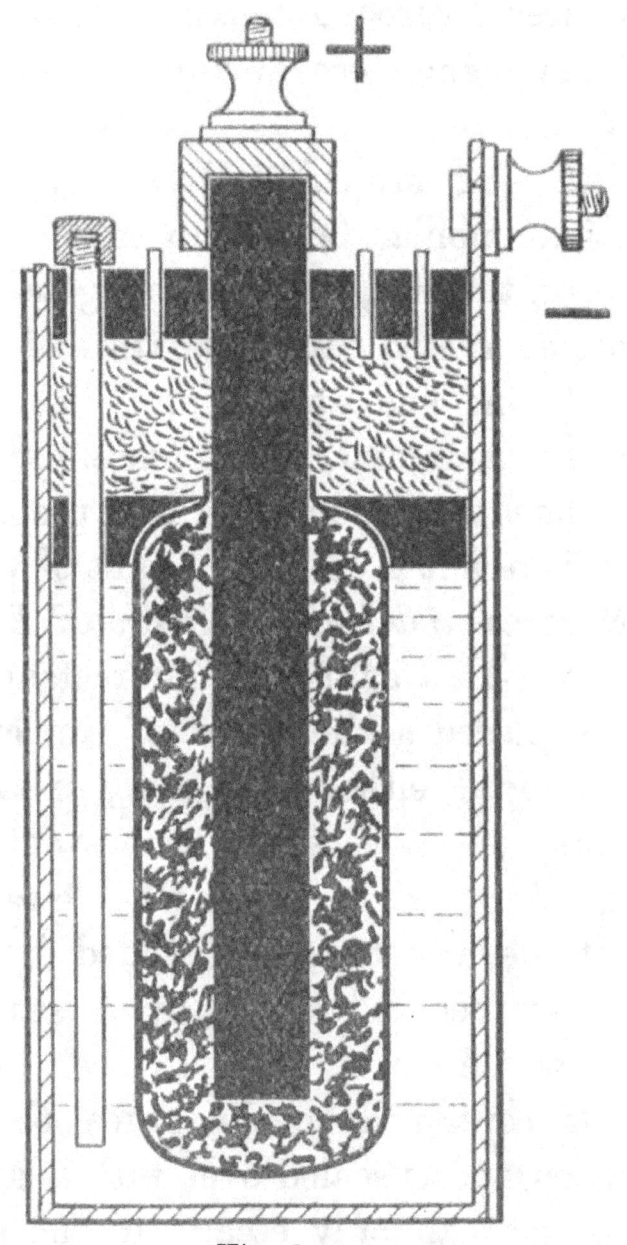

Fig. 84.

of liquid. A capped vulcanite tube extending to lower chamber and cemented into the sealing serves for the introduction of solution. Fig. 84 illustrates this cell.

§ 138. *Herr F. Pfannenberg's "Reform Cell."* Constituents:—Zinc and carbon, in sal ammoniac solution. Arrangement:—Two carbons are united at the top, by a bridge piece, and each is encased in a depolarizing mixture. The zinc is a long strip which is bent round outside the carbons, while the centre is turned in so as to form an *S*, the middle of which acts like a partition between them. The zinc is supported between two insulating supports, which prevent it from coming into contact with the carbons or the bottom of the cell. Herr Zacharias has subjected the ordinary size (20 centimetres) of this cell to severe tests.

Starting E.M.F. 1·48 volts. Starting current 1·1 amps.
After 12 hours ,, 1·07 ,, After 12 hours current 1·7 to 1·07 amps.
,, 20 ,, ,, 1·01 ,, ,, 20 ,, ,, ·65 amps.

Cells having an electrolyte containing calcium chloride gave better results than those with ammonium chloride alone.

§ 139. *Tyer's Pyrolusite[1] Cell.* Constituents:—Zinc in sal ammoniac, and carbon in manganese dioxide. Arrangement:—The cell consists of a porcelain vessel divided into two unequal portions by a porcelain partition, perforated all over. In the larger compartment is placed the zinc and ammonium chloride solution. In the smaller portion is the carbon plate and manganese dioxide mixture. The cell quickly polarizes. The "permanent cell" by *Marcus* belongs to this group of cells. Further modifications have been brought out by Clark, Muirhead, Binder, Gaiffe, Lieter, and others.

[1] Pyrolusite is the name of a natural manganese dioxide mineral.

§ 140. *Lieter's Portable Pyrolusite Cell.* Constituents :—
Same as Leclanché. Arrangement :—Outer cell made of
gutta-percha, in which stands a central perforated gutta-
percha cylinder containing a zinc rod. The carbon rod is
packed in the space between the outer cell and the
cylinder, with the usual carbon and manganese mixture.
A platinum wire connects the carbon to a piece of zinc
acting as a terminal. A solution of ammonium chloride is
used. The carbon and packing are covered with asphalte.
Said to be cheap, easily managed, and to give a fairly
constant and strong current. Used for medical purposes.
Except in the material for the case, and the use of
platinum connection for the carbon, it presents no point
of difference from the ordinary Leclanché.

§ 141. *The Trough Leclanché.* Arrangement :—A teak
trough divided by porous partitions, and coated throughout with marine glue. Lead-capped carbon and zinc
plates are placed in alternate cells. The carbons are
surrounded with the usual mixture, and the zincs are in
sal ammoniac solution. The zinc and carbon plates are
connected in pairs by means of an iron wire welded into
the zinc and the lead tops of the carbon. This form
dispenses entirely with glass cells; the porous partitions
are not so liable to crack as pots are, and the short length
of connecting wire, especially if thoroughly protected,
reduces to a minimum any danger of short circuiting with
it. If the zincs are scraped clean, and the solution of
sal ammoniac renewed every three months, the battery will
remain in action for years.

§ 142. Under the general name of "*Sack*" cells, several

manufacturers have placed on the market modifications of
the Leclanché, which are closely allied to the Siemens-
Halske cell described at § 137. These, owing to their low
internal resistance, are convenient for supplying fairly
heavy currents for short intervals; but they all require a
considerable amount of *rest*, after each discharge, in order
to allow them to recuperate, as depolarization does not

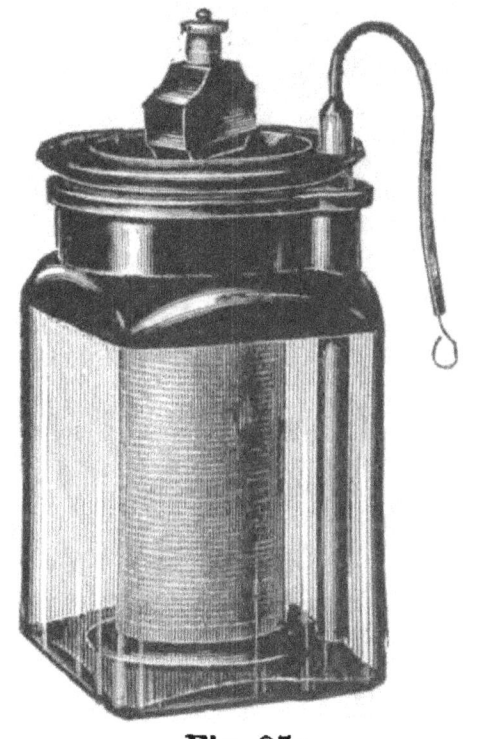

Fig. 85*a*.

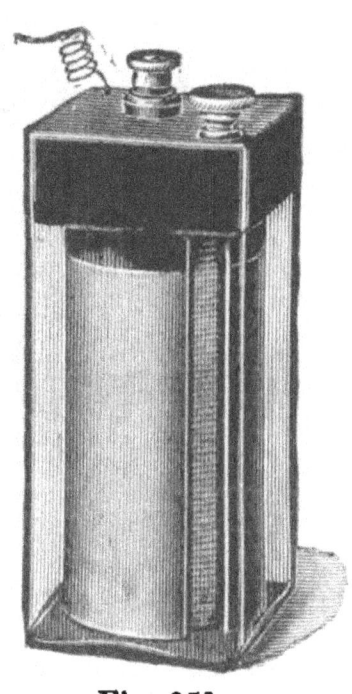

Fig. 85*b*.

take place very quickly. As examples of these, we may
mention the "*Sack* Leclanché" of the General Electric
Co., of which we give three illustrations at Fig. 85, *a*, *b*
and *c*. In this we have a central carbon rod or plate surrounded by the ordinary depolarizing mixture of granular
carbon and manganese dioxide, the whole being retained
in a· canvas sack, having a closed white ·porcelain top
and base. In the larger size, *a*, suitable for use with

Q

large electric bells, the zinc is a well-amalgamated rod standing outside the sack; in the medium size, *b*, the zinc is a cylindrical sheet, reaching to the bottom of the cell; while in the smallest cell, *c*, which is intended for more constant current, the zinc cylinder reaches only half-way down the cell; the zinc salt formed, falling to the bottom of the cell, owing to its greater

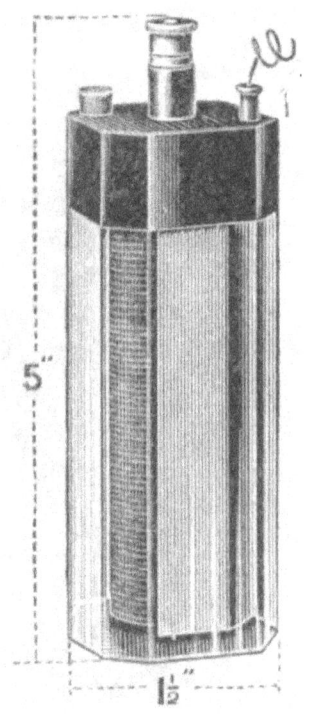
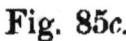

Fig. 85*c*.

Fig. 86.

specific gravity. The excitant recommended by this firm, is their "Salectron" in the proportions of 6 ozs. to the pint of water. Sal ammoniac may be used instead; but the firm claims that a cell charged with salectron lasts twice as long as when sal ammoniac is used, while the zinc is consumed in a much more even manner, no pitting or formation of zinc oxychloride being observed.

§ 143. The French firm "Le Carbone" (late Lacombe

and Cie.) turn out sack batteries very similar if not absolutely identical with the above. Our Fig. 86 shows one of their "sacks," containing the carbon, and the depolarizer; Fig. 87 presents a view of the internal arrangement of a mounted cell containing the sack and

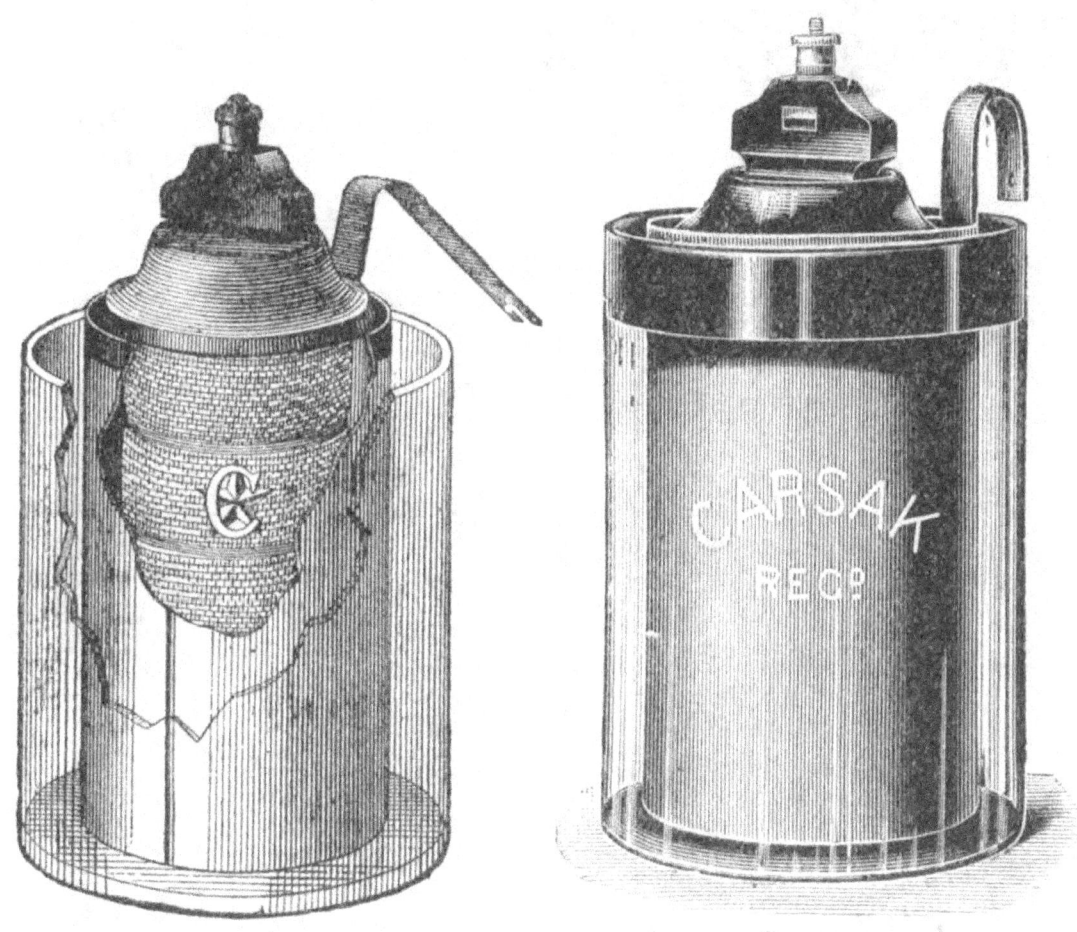

Fig. 87.   Fig. 88.

the surrounding zinc cylinder. The "Carsak" of the General Electric Co., Fig. 88, and the "Z" cell of "Le Carbone" Co. are almost identical, and an ingenious modification of the sack form of porous cell, adapted to special requirements.

§ 144. *Lacombe* of Paris, some ten or twelve years ago,

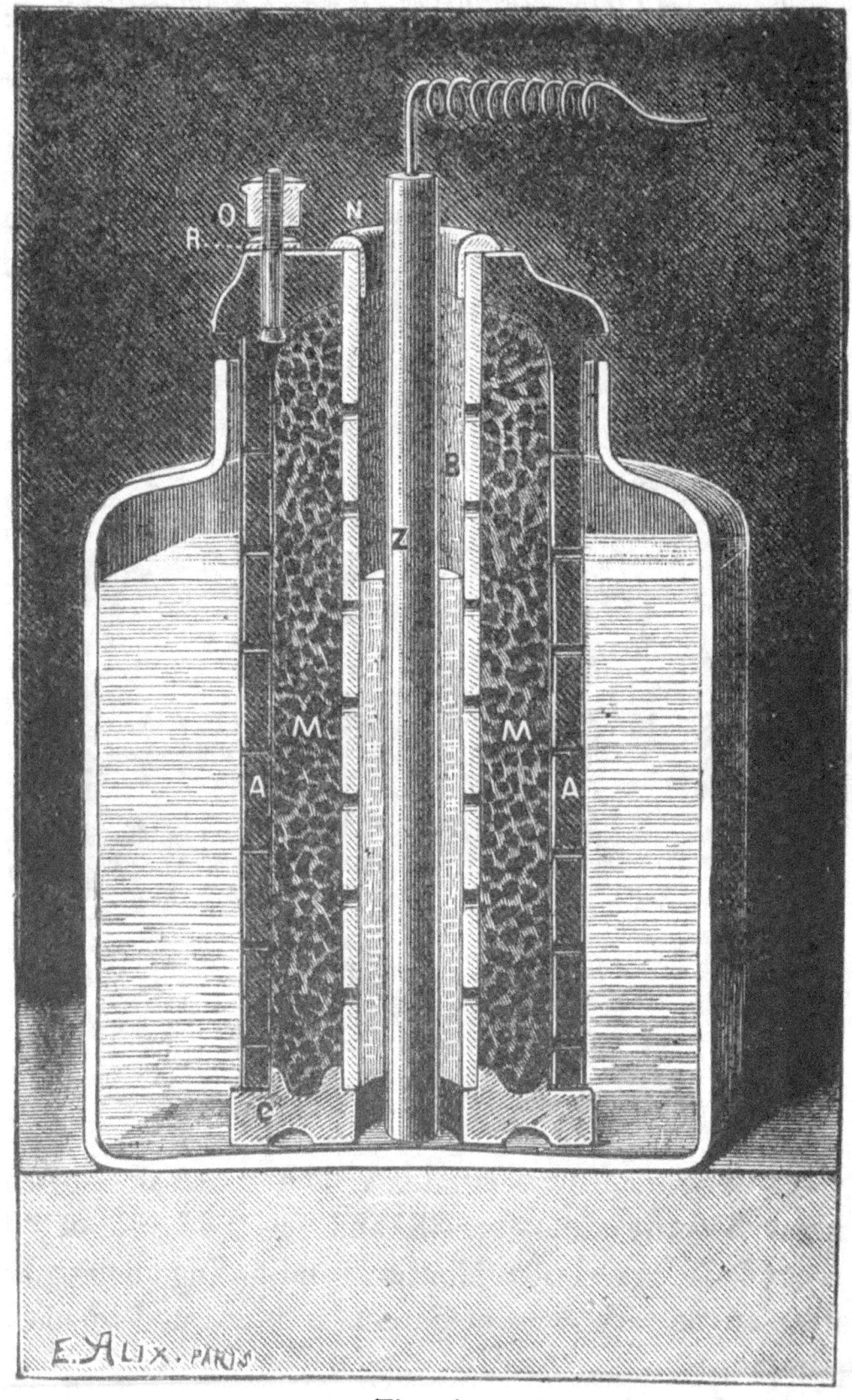

Fig. 89.

devised a very elegant and efficient form of Leclanché known in England as the " Central zinc " or " Carporous " cell. Both the General Electric Co. and Le Carbone Co. supply this cell. Fig. 89 is a sectional view of a cell of this type. *A* is a perforated carbon cylinder with head ; *B* a cylinder of porous ware pierced with many holes ; *C* is a glass foot or base, uniting together the two cylinders

Fig. 90.

A and *B*. *M M* is an annular space between the said two cylinders, containing depolarizing mixture. The zinc rod *Z* is placed in the centre of the porous cylinder *B*, and is thus surrounded by the depolarizing mixture. *N* is an insulating flange ; *O* the screw contact terminal, running down to *R*, which is a washer ensuring contact. Fig. 90 illustrates the same cell made up. This form of central zinc cell lends itself readily to use with other depolarizers besides manganese dioxide; in fact, by the addition of dilute sulphuric acid to the excitant, along with about 5 per cent of chromic acid, a very excellent result is obtained, the E.M.F. rising to 2 volts. As the internal resistance is very low, owing to the large carbon surface, a heavy current may be taken off for a considerable time; and the depolarization takes place quickly, which is due to the presence of both chromic acid and manganese dioxide (see also Duffett's cell, § 175, which is an almost similar arrangement).

§ 145. *Messrs. Mix and Genest* of the International Electric Co. have introduced a modification of the Leclanché, to which they have given the name "Manganese" cell, and which may be looked upon as a hybrid between the ordinary Leclanché and the Agglomerate (*q. v.*). They state:—"Experience proves that the ordinary Leclanché cell soon becomes exhausted on account of the short circuiting which is caused by the zinc reaching to the bottom of the cell. The chloride of zinc formed during the work of the cell sinks, on account of its higher specific gravity, to the bottom of the cell, displacing the sal ammoniac solution, which rises to the top. As the two liquids have different conductivities, and the electrodes are inserted in both, a current, starting from the zinc, passes through the better conducting fluid to the carbon, and thence through the liquid of lesser conductivity back to the zinc, even when the cell is not doing useful work, thus exhausting it much more rapidly than would be the case if no such action took place. In order to overcome this defect, we have devised a new form of Manganese battery, in which the rod of zinc is replaced by a *cylinder* reaching about two-thirds only of the way down the cell, which cylinder is supported by lugs resting on the edge of the jar (see Fig. 91). A smaller cylinder, consisting of a compressed mass of carbon and manganese dioxide, is placed upright in the centre of the cell, and being

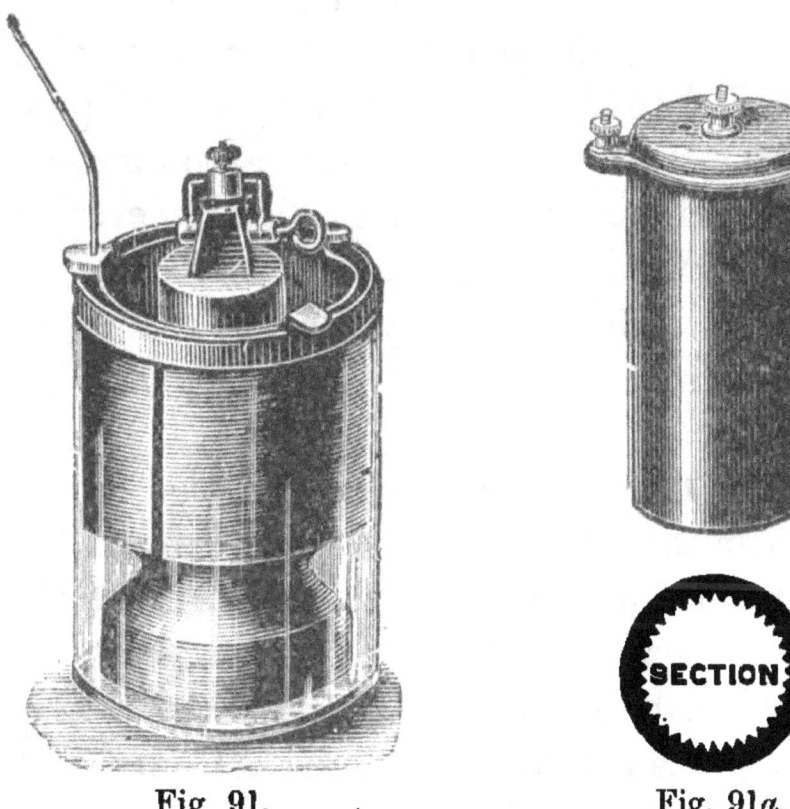

Fig 91.   Fig. 91*a*.

provided with a large circular foot, of the same material, stands without any support, or porous cell, to prevent it touching the zinc." With this construction the zinc chloride sinks to the bottom of the cell, as soon as it is formed, and is thus out of contact with the zinc. Like the other cells of this class, the E.M.F. is about 1·5 volts, when freshly made up; and the internal resistance of a quart cell is about ·6 of an ohm. *Applegarth*, in 1878, patented a process of fluting or grooving carbon plates or cylinders, with a view to increasing the superficies. Shortly afterwards *Judson* brought out a form of Leclanché, in which a cylinder of inwardly corrugated carbon is encased in an outer coating of insulating composition. In the cell are two or more thin carbon plates, full of perforations, these plates being cemented to the sides of the cell by Prout's elastic glue, so as to leave spaces between the plates and the sides of the cell; these spaces being filled in with the usual manganese carbon mixture. The zinc rod, which is affixed to an insulating cover, stands in the centre of the cell, touching it at no part. As usual, the exciting fluid is the ordinary half-saturate sal ammoniac solution. Fig. 91*a* gives a good idea of the Judson cell.

§ 146. To the Leclanché class belongs *Howell's* cell. It consists in the usual zinc rod and carbon. The latter is placed outside the porous pot along with carbon granules, manganese dioxide and manganese sulphate, and covered with dilute sulphuric acid. The zinc rod in the porous pot is immersed in a 2½ per cent. solution of ammonium sulphate. It is claimed by the inventor, that this cell gives an E.M.F. of 2·25 volts, with an internal resistance of from 5 to 6 ohms. The latter figures may be correct, though they are rather high for a carbon-packed cell; but the voltage stated is certainly wrong: it does not exceed 1·6 volts.

§ 147. *Pollack's* " Recuperative Gravity Leclanché " (1886) differs from the ordinary type in general arrangement in having copper electrolytically deposited on the lower extremity of the carbon. The arrangement is as shown in our Fig. 92, wherein *Zn* is a massive ring of zinc, lying at the bottom of the cell, and passing up through an insulating tube to the terminal. The cell is then half filled with a solution of sal ammoniac in water, strength about 6 oz. to the pint. A wooden cover holds,

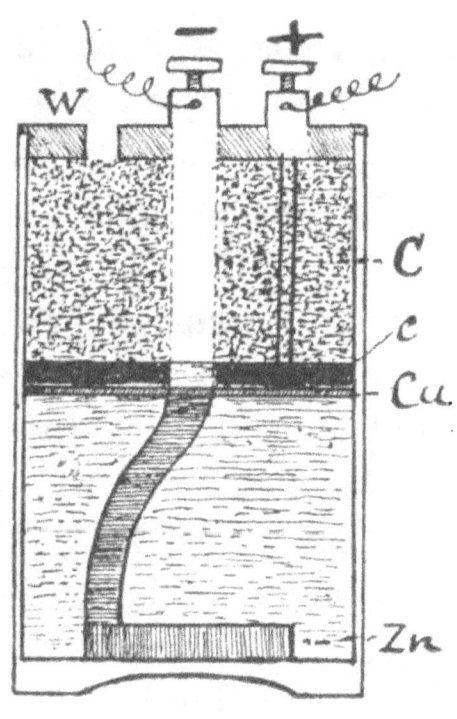

Fig. 92.

by means of a screwed rod, a carbon disc *c* just below the level of the liquid. The under part of this disc is heavily coppered. Above the carbon disc *c* the space between it and the cover is packed tightly with granular carbon. The first reaction which takes place is precisely similar to that of an ordinary Leclanché, but in the meanwhile, the sal ammoniac solution is acted on by the carbon-copper couple, with the production of copper chloride. The hydrogen evolved during the action of the battery decomposes the copper chloride, re-precipitating the copper on the carbon disc, so that the cell is practically regenerative, as far as the depolarizer is concerned. E.M.F. about 1·3 volts: I.R. 0·5 ohm.

§ 148. *Clamond and Gaiffe's* Leclanché differs from the typical form only in the depolarizer, which is sesquioxide of iron, instead of manganese dioxide. E.M.F. 1·2 volts, internal resistance high, about 5 ohms in the quart size.

The American cell "Axo Leclanché" has the porous cell flanged at the top, to form a cover to cell. The carbon rod has longitudinal furrows, to facilitate the escape of the hydrogen.

§ 149. A form of the Leclanché, partaking of the *dry cell* character, is the *Silvertown*. Two carbon plates, capped with lead, and connected together both above and below by lead strips, have inserted between them a zinc plate enclosed in a coarse flannel bag. This bag is filled with crushed sal ammoniac. The whole is put in any water-tight case, packed with crushed carbon and manganese dioxide, and moistened with sal ammoniac solution, the whole being sealed (except at the terminals and a vent-hole) with melted pitch. When intended for blasting purposes (for which it has been extensively used by the Royal Engineers) it was slightly modified in arrangement. The outer case was given a square shape, the zinc sheet was bent into the form of a ∪, between the limbs of which was placed the carbon plate contained in a canvas bag rammed full of crushed carbon and manganese dioxide. Excitant as before, a strong solution of sal ammoniac. This cell, like the former, was sealed up with pitch. Owing to the large surfaces, both of zinc and of carbon, this form of cell was capable of giving at intervals a large "splash" of current, suitable for fuse-firing; after which the current fell to a mere nothing, until time had been given for it to recover.

Holtzer's Leclanché consists in a carbon cylinder with a flanged foot. In the centre of this stands the zinc rod, which is prevented from coming into actual contact with the carbon by means of one or more porcelain rings encircling it. The excitant is the usual sal ammoniac solution. No depolarizer is employed. Described in the *English Mechanic*, May 31, 1889.

Hayden's "Perforated carbon cell" is almost precisely the same as the Lacombe, *q. v.* Described in 1894.

§ 150. *Niaudet*, in 1879, devised a modification of the Leclanché, in which the manganese dioxide was replaced by fragments of chloride of lime. The depolarizing agent is here the chlorine gradually evolved. The best excitant for this cell is a solution of common salt; though sal ammoniac may be used without any serious detriment. The initial E.M.F. is 1·65 volts. After some months this falls to 1·5 volts. The internal resistance of the pint size about 0·5 ohm. Action takes place when the circuit is closed only. The smell is to some disagreeable, so that the cells should be hermetically sealed.

§ 151. The *"Insulite"* cell (1883) is a square box of ebonite, divided down the centre diagonally by a porous partition. On one side of this partition is a carbon rod packed with carbon granules; on the other side is a rod or plate of zinc. The two divisions are filled nearly to the top with a mixture of sal ammoniac and common salt in water, after which a cover conveying the terminals is fitted on the top of the case, and cemented thereto with a pitchy composition. It is therefore, practically, a sealed Leclanché. E.M.F. 1·6 volts. Internal resistance, 1·5 ohms.

§ 152. *Binko's* "Electric bell battery" (1882) is simply an agglomerate cell, in which the agglomerate blocks contain a mixture of metallic oxides. The composition consists of 50 parts carbon and 12½ parts each of pyrolusite, of iron filings, iron peroxide, and copper dioxide, moulded into blocks under heavy pressure. The zinc is either held near the blocks by rubber bands, or is contained in a small porous cell. The exciting fluid is a half-saturate solution of sodium or ammonium chloride. E.M.F. 1·6. Said to be very permanent.

§ 153. We called attention to the Grove cell at § 50, and refer the reader to p. 95 for description and illustrations. In order to render clear the direction in which modifications have been introduced, we summarize briefly its constructional details. The Grove cell consists essentially of a platinum sheet standing in a flat porous cell, nearly full of strong nitric acid, which porous cell is placed in an outer containing vessel, in which is placed a sheet of zinc bent into the shape of a ∪, so as to embrace the porous cell; the exciting fluid along with the zinc being dilute sulphuric acid. The zinc should of course be well amalgamated. The E.M.F. of the Grove cell is 1·96 volts; and owing to the comparatively low resistance of the nitric acid employed, it gives a large current: eight ampères being readily obtained from a quart-size cell. Its defects are threefold : 1st, the high price of platinum ; 2nd, the generation of noxious and corrosive fumes of nitrous acid ; 3rd, the excessive consumption of zinc. The following gives an idea of the suggestions made and carried out by several experimentalists, to overcome some or all of these defects.

§ 154. *Cliff's* cell (1870) is a modification of the Grove, in which a successful attempt has been made to retain the activity of the nitric acid without the inconvenience of the nitrous fumes. A rather long cylindrical stoneware jar is fitted with a central porous pot, rather shorter than itself. Over this is arranged a well-fitting earthenware cover, of the shape of an ordinary saucer, the edges of which rest on the outer jar, the rim at the bottom fitting accurately in the porous cell. In the centre of this saucer are seven holes, one central, through which the extension of the platinum passes, and six lateral, to allow the escape of fumes. The zinc is in the form of a cylinder, with lug which passes out through a recess in the under part of saucer, standing between the outer jar and the porous pot. The charge is, as usual, dilute sulphuric acid along with the zinc, and strong nitric in the porous pot, along with the platinum. The peculiarity consists in the saucer-like cover. This is filled with small lumps of freshly-slaked lime, which effectually prevent the evolution of nitrous vapours. The lime naturally requires renewing from time to time, when it has become saturated with nitrous vapour. Fig. 93 gives a sectional view of this cell.

§ 155. *Bunsen*, in 1840, used a carbon cylinder instead of a platinum sheet, placing it *outside* the porous pot, in which stood the zinc. The exciting fluid, as also the

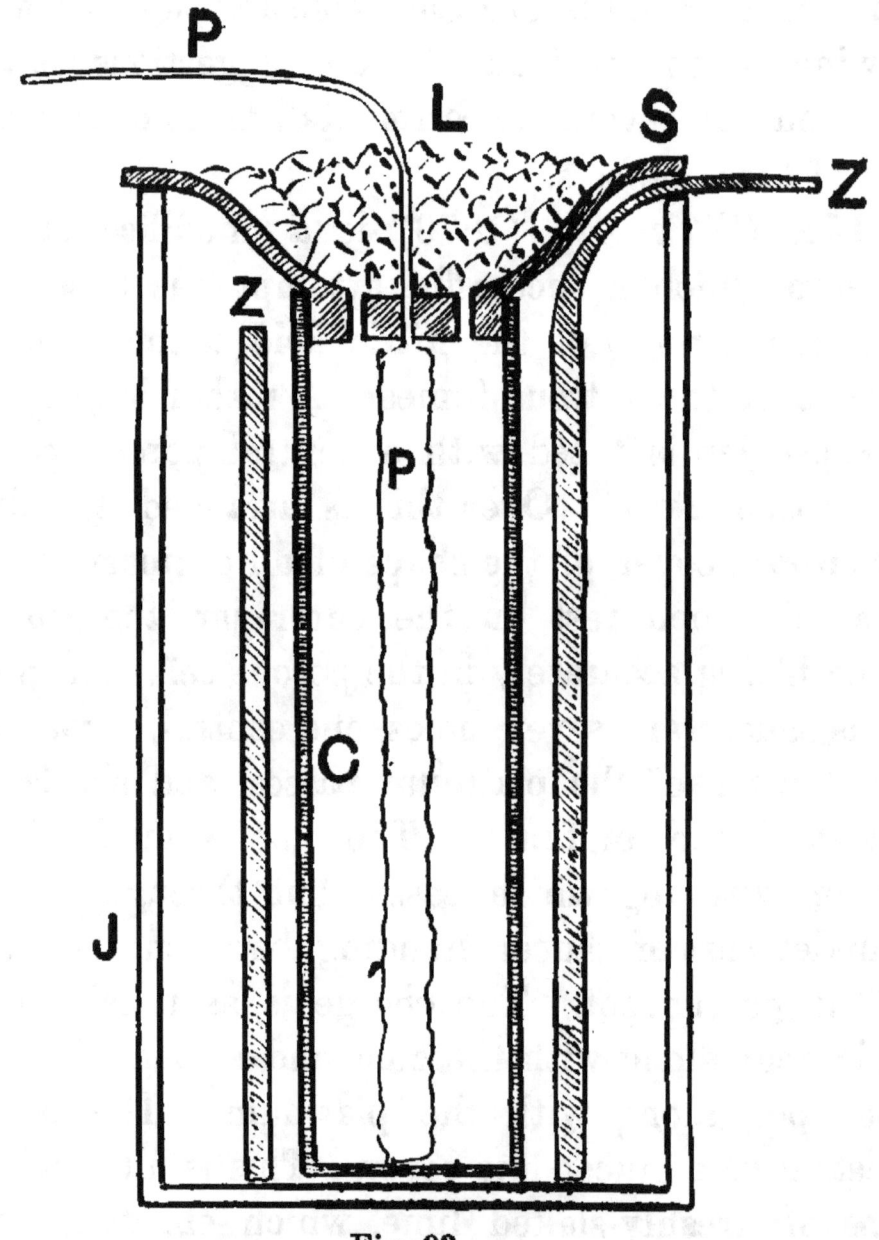

Fig. 93.

depolarizer to be employed in such a cell, are precisely as in the original Grove, viz. dilute sulphuric along with the zinc, and strong commercial nitric acid in the carbon compartment. Owing to the large negative surface exposed, due to the roughness of the carbon, the Bunsen is, size for size, rather more efficient than the Grove. Like the Grove, it can only be used out of doors, or in situations provided with a proper "stink cupboard," to carry off the injurious and offensive nitrous fumes. E.M.F. 1·96 volts; internal resistance of one quart size, 0·08 ohm; so that

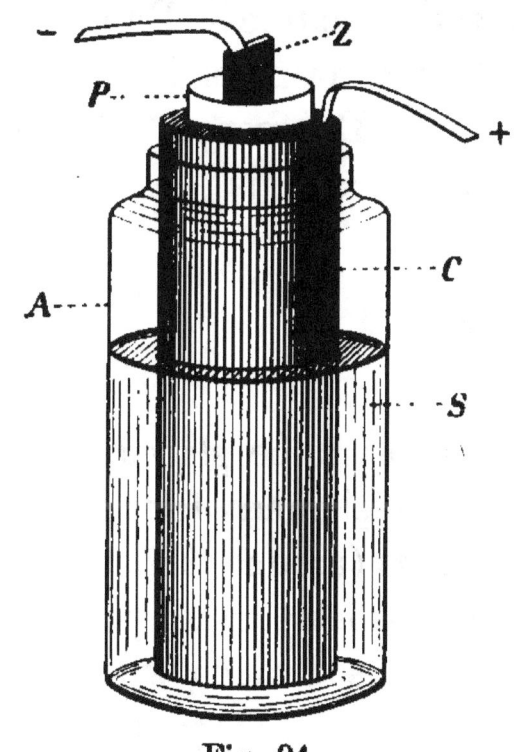

Fig. 94.

about 24 ampères can be got on a short circuit from a quart-sized cell. One disadvantage in the Bunsen type of cell, is the rapidity with which the nitric acid is sucked up by the carbon cylinder, to the detriment of the connecting terminals. At Fig. 94 we illustrate the Bunsen cell.

§ 156. *Archereau* in 1842 modified the Bunsen to its

present form, by using a square or round graphite block in the porous cell, along with the nitric acid, while the zinc takes the form of an outer cylinder. Except as a matter of ease in making, such a construction is not an advantage, since the larger negative surface presented by a cylinder of carbon, as compared to a central block, enables the cell, owing to lower internal resistance, to give a larger current.

§ 157. *D'Arsonval's* modified Bunsen (1879) differs from the standard type, principally in the nature of the depolarizer employed. D'Arsonval's formula for this, to be used instead of nitric acid, is—

Nitric acid ...	...	...	...	1 part
Hydrochloric acid ...	...	...	1 „
Water	...	...	...	...	2 parts

He recommends the zinc to be placed in a central porous pot, with the negative element formed of a "crown" or cylinder of carbon rods (around the upper extremities of which is cast a type-metal ring) surrounding it. The exciting solution to be used along with the zinc is dilute sulphuric acid, consisting of 1 part oil of vitriol to 20 parts, by measure, of water. It is claimed for this modification, that the E.M.F. is higher, and the internal resistance lower than with the ordinary nitric acid depolarizer, and carbon plates or blocks; so that a cell 4″ in height will give as much as 40 ampères on the short circuit. This would imply that the E.M.F. being 2 volts, the internal resistance is only 0·05, instead of being 0·08 to 0·11 as in the ordinary Bunsen.

§ 158. *Magnesium Bunsen* cell. Consists in a rod of magnesium standing in a porous cell filled with sodium silicate solution (water glass) surrounded by a carbon cylinder standing in a dilute solution of sulphuric acid, along with sodium bichromate and a little sodium permanganate. E.M.F. very high; said to be 3·5 volts.

In the '80's, many inventors such as *Holmes and Burke, Burr, Ross, Webster*, etc., claimed to have produced cheap modifications of the Bunsen cell, by using a mixture of nitrate of soda and strong sulphuric acid, instead of nitric acid in the carbon compartment. But very little economy if any is obtained. The expense of a cell chiefly lies in the zinc consumed and the labour in setting up. The attempts made to utilize the waste products have not hitherto been encouraging. For instance, the sale of the " black mud " (metallic copper) produced in 70,000 cells of Daniell during one year, realized only £167 14s., or a fraction over a halfpenny per cell.

The modification imported into the Bunsen by *Tweedale* is very slight. It consists in the employment of mixed nitric and sulphuric acid in the porous cell along with the carbon, instead of nitric acid only. It is doubtful whether this addition is any real improvement. Although its inventor seemed to think highly of it, when he announced the construction of it in 1894, it appears never to have met with much favour.

§ 159. Efforts have been made, hitherto without any great success, to find a cheap substitute for zinc in those cells. *Mn. Rousse* proposed lead or iron attacked by nitric acid. *M. Maiche* in 1864 used a cylinder of sheet-iron,

R

acted on by water containing one per cent. of strong nitric acid. *Callan, Hawkins* 1840, and *Schönbein*, 1842, having shown that cast- or sheet-iron which has been immersed for a few moments in strong nitric acid and then washed, is "passive" or electro-negative to another piece of ordinary iron, when plunged into ordinary exciting solution, proposed and gave their names to modifications of the Bunsen cell in which the carbon is replaced by "passive" iron. But these cells are not trustworthy; very often, from some little understood cause, the nitric acid will violently attack the "passive" iron, and the fluid will boil up suddenly, with copious evolution of the red fumes of nitrous oxide. Attempts have also been made to replace the nitric acid in the Bunsen, by other and less objectionable depolarizers, such as chlorate of potash (Leeson, 1843), chloric acid, perchloride of iron, picric acid, chromic acid, or bichromate of potash (Bunsen, 1843). This latter leads us to the considerations of the two-fluid "chromic acid" or "bichromate" type.

§ 160. In the modified form, as suggested by Bunsen, the zinc rod, standing in the porous pot, was excited by dilute sulphuric acid (1 to 12). The depolarizer was a saturated solution of bichromate of potash, to which $\frac{1}{8}$ of its bulk of strong oil of vitriol was added : the mixture being allowed to cool before use. This forms an excellent working battery, and one which has undergone many changes of name at the hands of different experimenters. The real improvements are few. For slow work *Fuller* suggested painting the porous pot (which contains the zinc) all over with paraffin wax, except along one narrow strip reaching from top to bottom, along one side. Rain-water is used for the excitant. By endosmose, sufficient mixed sulphuric acid and bichromate permeates from the carbon compartment to act upon the zinc. (See Fig. 95.) *N. H. Warren's* "Boron" cell is simply a double-fluid bichromate. It is neither better nor worse than others of its class. Some profess to believe that the plates con-

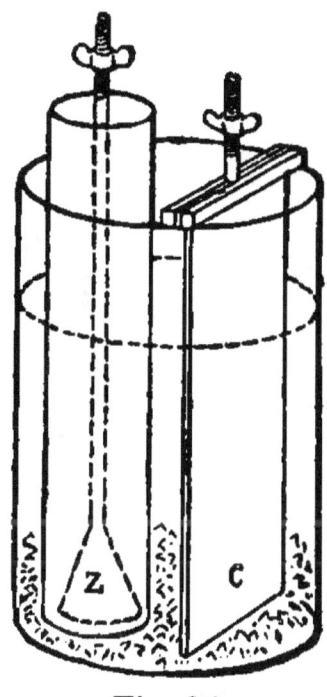

Fig. 95.

tain boron along with the carbon; but even if this be the case, it does not affect the efficiency of the cell. The employment of *chromic acid* in these double-fluid cells, especially in conjunction with a little potassium chlorate, greatly increases the output and lasting powers.

§ 161. *Anderson's Double-Fluid Cell.* Here we have a zinc rod standing in a solution of ammonium chloride contained in a porous cell, outside this is the carbon plate or cylinder immersed in the outer vessel which contains a solution of double oxalate of potassium and chromium, prepared as described at § 94; or, if very energetic action be required, the carbon compartment is fitted as shown at our Fig. 96, with a wide-mouthed glass funnel having a perforated bottom; into this funnel are placed crystals of bichromate of potash, and hydrochloric acid diluted with

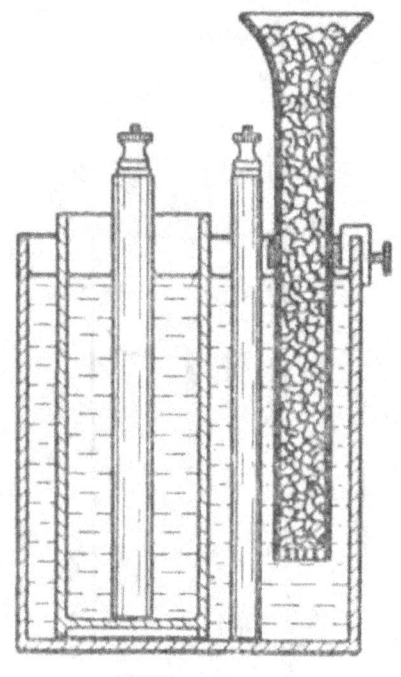

Fig. 96.

an equal volume of water is poured in until the entire vessel is nearly full. By raising or lowering the funnel, so as to permit or to prevent more of the strong bichromate solution passing into the cell, the action of the cell can be regulated to a nicety.

§ 162. *Dale's Zinc Chloride Cell* (1881). Has for elements amalgamated zinc and carbon. The zinc stands in a solution of 1 part of chloride of zinc in 20 parts of water, placed in a porous cell. The carbon (plate or cylinder) is contained

in an outer vessel, in which is the depolarizer, consisting
of 2½ parts of bichromate of potash, 2 parts of hydrochloric
acid, and 20 parts of water. It is said there is no action
on the open circuit. This cell is fairly constant; but the
hydrochloric acid fumes are objectionable (see Fig. 97).

§ 163. In *Julius Thomsen's* constant copper battery we

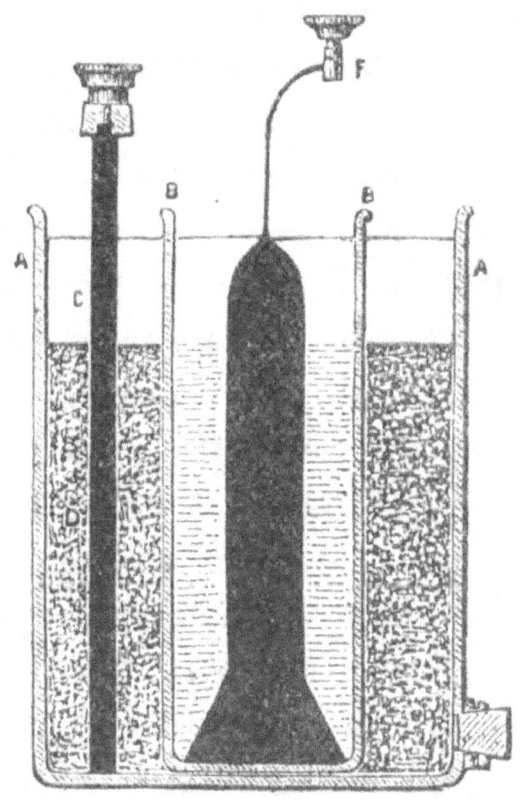

Fig. 97.

have a new departure. The active metal is copper, which
stands in a solution of sulphuric acid, 1 part of acid to 4
parts of water, contained in a porous cell. Outside this
is the carbon element with its depolarizer, consisting of
25 parts sulphuric acid, 12 parts of bichromate of potash,
and 100 parts of water. The copper is not attacked on
open circuit, and the internal resistance is very low.

§ 164. *Mr. Moleyris* in 1868 devised an extraordinary

form of cell in which the negative element is practically a slab of boxwood charred on the surface; this stands in a solution of 6 ounces of nitrate of ammonia dissolved in 2 ounces of water, to which is added an equal bulk of strong sulphuric acid. The zinc, which need not be amalgamated, stands in a saturated solution of ammonium chloride. One peculiarity, on which the inventor lays great stress, is keeping the sulphuric acid mixture at a temperature of below 100° Fahr., by means of a frigorific mixture!!

§ 165. *D. G. Fitzgerald* arranged a calcium chromate cell consisting of a rectangular outer vessel divided into water-tight compartments by means of a carbon plate. This plate is perforated with small holes, which are plugged with porous earthenware. On one side of this carbon is placed a plate of amalgamated zinc, acted on by 1 part of sulphuric acid in 10 of water; the other side is filled in with a mixture of equal parts of calcium chromate, water and sulphuric acid. Said to be very energetic in action.

§ 166. *Gravity Bichromate Cell.* Constituents: Carbon and amalgamated zinc in sawdust moistened with bichromate solution. Arrangement:—An outer cell 15″ deep and 5″ outside diameter, having an opening in the bottom, capable of letting out 1 pint of any liquid in the cell. First is placed a small piece of round wire gauze over the opening; then is lightly put in a thin layer of fine sawdust about ¼″ thick. Over this is placed a carbon plate, having a gutta-percha-covered wire connection. This is followed by a layer of sawdust, on which is laid a zinc plate, also furnished with insulated wire connection. Proceeding thus for the required number of plates, bringing up all the connecting wires to two separate terminals, the cell is filled to within about 4″ of the top with a solution of 2 ounces of potassium bichromate in 1 pint of water, adding 1 ounce of strong sulphuric acid to every pint; and more when required.

§ 167. *Luigi Ponci's* two-fluid bichromate consists in zinc and carbon elements, in a solution of 1 kilogram of bichromate of potash, 4 litres of water, and 2 litres of hydrochloric acid (or 5 ounces of bichromate of potash, 1 pint of water, and ½ pint of hydrochloric acid). The bichromate is crushed and dissolved in boiling water, and after complete solution, all the acid is added. The advantage is that no crystals are formed by the working of the cell. Arrangement:—Six rectangular glass cells, whose bottoms are pierced, rest in a lead-lined wooden case on glass strips, so that the liquid can circulate throughout the battery. Each glass vessel contains a zinc plate with a carbon plate on each side, sandwich-fashion. The case has a lead syphon for filling and emptying the battery. A battery of 12 cells (2 cases) will do the work of 6 small-size Bunsen cells for two hours, and with 8 cases a good arc light can be produced.

§ 168. *Messrs. Oliphant, Burr and Gowan* introduced some time ago a two-fluid bichromate cell, in which the excitant was water, in which a little mercuric chloride was dissolved, the depolarizer being the usual potash bichromate solution. Oddly enough, they recommended the zinc to be lightly gilt!

§ 169. *Chloris Baudet*, in 1885, brought out his "Unpolarizable" cell. The constituents are zinc in an acid solution of sulphate of potash ($HK SO_4$) and carbon in a saturated solution of bichromate of potash along with sulphuric acid. The arrangement is: an outer cell about $8\frac{1}{2}''$ high by $7''$ long and $3\frac{1}{2}''$ wide contains the bichromate of potash solution. A porous cell stands in the centre, and this latter contains the zinc plate along with the acid sulphate of potash solution. On either side of this porous cell is a carbon plate. In order to maintain the bichromate solution at a constant strength, a small porous cell is placed on each side of the central one; of these, one contains a supply of bichromate crystals, the other is filled with strong sulphuric acid. Such a cell was found to give a steady current of rather over 1 ampère for 28 consecutive hours. E.M.F. 1·87 volts.

§ 170. In 1886 *Maquay* recommended a cell of his design, which he designated "the Regent." The active metal is an alloy of 95 parts of zinc, 2 parts of tin, 2 parts of lead, and 1 part of mercury, cast into plates or rods, and then well amalgamated. In a porous cell is placed a carbon or platinum plate, which has been dipped in sulphur to the depth of from $\frac{1}{20}$ to $\frac{1}{8}$ of its length: or to the same end, granules of sulphur may be put in the porous cell, along with the negative plate. The exciting fluid is a 5 per cent. solution of sulphuric acid in water; the depolarizer consists of potassium bichromate 40 parts, nitric acid 40 parts, strong sulphuric acid 20 parts. The outer cells are of glass or porcelain, and they communicate with one another at about $\frac{1}{2}''$ from the bottom, by holes, for convenience of emptying. These holes may be closed at will by plungers, which, together with the battery plates, are mounted on a top plate of ebonite. When this top plate is raised, all the positive cells communicate, the negative cells being likewise put into connection with one another. On lowering the top, all the openings are closed, so that the cells become independent of each other.

§ 171. In *Howard's* cell we have the first application of paraffin to prevent consumption of the zinc on those portions not facing the negative element. In this cell, a cylinder of zinc, paraffin waxed on the inside, is placed in a porous cell containing a 10 per cent. solution of sulphuric acid in water. Around this, in a suitable containing vessel, is a cylinder of carbon, standing in a solution of sodium bichromate and common salt in water. The solution should be saturated. This was the forerunner of the "Fuller" cell, and was made known in 1886.

§ 172. *Kousmine* (1890) took advantage of the great difference in specific gravity between sulphuric acid and bichromate solution, to construct a "gravity" form of cell. To this end he arranged, in a cylindical glass vessel 8″ high, 6″ in diameter, a circular zinc grating or grid, at the bottom. Four carbon strips, about 5″ long, connected together at the top, were attached to the cover. Sulphuric acid of the sp. g. of 15° Beaumé, was poured in, to the level of the lower ends of the carbons. Then a 6 per cent. solution of potassium bichromate was carefully added, so as not to mix with the subjacent sulphuric acid on which it floated. The bichromate solution is so weak that the chrome alum formed is all dissolved, thus preventing clogging of the carbons. If used on the short circuit, a violet ring of spent solution shows itself at the level of the lower ends of the carbons. The zinc sulphate formed falls to the bottom, causing the sulphuric acid to rise. This cell is another attempt to prevent the increase of internal resistance, by setting up gradual diffusion. It has been much used in Russia. Tried before a committee of experts, after running on an outer resistance of 0·32 ohm for 8½ hours, and then left on the open circuit for another 10½ hours, it continued to work for another 4½ hours, giving a total of 36 ampère hours for 13 hours with a consumption of 48 grams (nearly 1·7 ounces of zinc).

§ 173. Under the somewhat fanciful name of *Osbo Premier*, a cell was put in the market in 1891 for the special purpose of electric lighting. The cell itself consisted in a carbon-zinc couple made on the single-fluid type with a special depolarizer along with dilute sulphuric acid. (The depolarizer may be a weak bichromate or chromic acid solution.) Besides the elements, each cell contained a porous earthenware tube with small perforations at the bottom, fitted with a stopper at the top. When the battery or cell was to be brought into action, a small bag containing a mixture of chloride of lime along with nitrate of nickel was inserted in the said tube, which was then carefully closed with the stopper. As soon as the liquid reached the chloride of lime mixture, oxygen was freely given off, which quickly revivified the depolarizer. E.M.F. 2 volts; current in a cell $+18'' +10'' +2''$, 14 ampères constant for 30 hours = 420 ampère hours.

§ 174. In the *Dowse's Cell* we have zinc in a solution of sal ammoniac, chromic acid, mercury bisulphate, and hydrochloric acid. The inventor says these two latter are not essential, the action being less intense but more even without them. Of course in the former case although the E.M.F. is higher, and current given larger, it is not so lasting. The other electrode is copper in copper sulphate. The arrangement is as follows :—An outer glass vessel made in any suitable form is fitted with convenient number, say 4, of porous cells, which are placed inside. Each porous cell contains a zinc electrode, preferably cruciform, so as to expose a large surface. The zincs are all joined together by an external metallic ring; the solution in the cells is the one given above. Outside the porous cells is bent a sheet of copper to fit the inside of the glass cell, which is then filled with a strong solution of sulphate of copper. Between the porous cells is placed a glass funnel dipping into copper sulphate solution, and this is filled with crystals of sulphate of copper which gradually dissolve as action takes place. The E.M.F. is nearly 2 volts, and owing to the large surface the internal resistance is low, so that a large current, say 4 or 5 ampères, may be readily drawn off. Manganese dioxide may be mixed with the chromic acid mixture above described, but this while conducing to the lasting properties of the cell also increases the internal resistance, and therefore lowers the current.

§ 175. *Duffett* of New Zealand, 1890, constructed a cell on the Leclanché type, in which the zinc stands in a porous pot containing sal ammoniac solution, around which is a perforated carbon cylinder surrounded by a paste of crushed carbon, peroxide of manganese, chromic and nitric acids. E.M.F. 2 volts, internal resistance in quart size about 0·5 ohm.

§ 176. In 1893 *Williams* constructed his "Excelsior" cell. This is a double-fluid zinc carbon battery, with a special "regenerative" depolarizer, made as follows: 3 parts of potassium permanganate, 3 parts of sodium nitrate, 8 parts of manganese dioxide, 2 parts of chlorochromic acid, and 20 parts of finely-crushed carbon are gently heated together, to incipient fusion. At this point a stream of oxygen gas is passed through the mixture, which is then allowed to cool, and is broken up into small pieces. The result is said to be a depolarizer which, when spent, is capable of re-absorbing oxygen from the atmospheric air. The zinc rod, previously well amalgamated by standing for two or three hours in an acid solution of mercuric nitrate, is placed in a porous pot with mercury at the bottom, along with a 20 per cent. solution of sulphuric acid in water. The carbon plate is embedded in the special depolarizer above described, moistened with dilute sulphuric acid. A flexible tube passes through the cover of the carbon compartment, and reaches to the bottom of the carbon. By injecting air into the cell through this tube, the depolarizer can be re-oxidized when spent. E.M.F. 1·785. Can give a current of 2 ampères for 12 hours. Size of cell, 6" high, 4" diameter.

§ 177. In *Edgar and Milburn's* two-fluid aluminium battery, the aluminium, either in the shape of rolled sheet or rod, stands in a large porous cell, containing a solution of 6 parts of ammonium chloride dissolved in 30 parts of water, to which are added 2 parts of hydrochloric acid. The porous cell stands in a glass or glazed outer vessel in which is inserted, either a carbon cylinder, or else a number of carbon plates joined together at the top so as to act as one plate. As a depolarizer, strong nitric acid is used in the carbon compartment. As aluminium is now so cheap, the arrangement of the two elements can be reversed to great advantage, and a carbon block may be stood in the porous cell and surrounded by nitric acid, while the aluminium, in the shape of a cylinder, may be placed *outside* the porous vessel, standing in the glass or glazed vessel, and dipping into the mixed ammonium chloride and hydrochloride acid solution. A mixture of 6 parts of potassium bisulphate in 15 parts of water has sometimes been recommended as a substitute for nitric acid in this cell: but this is evidently a mistake, since this salt does not act as a depolarizer; but if about an equal weight of soda nitrate be used *along* with the bisulphate, better results are obtained.

§ 178. *Sturgeon* in 1840, followed by *Callan, Maynooth, Turton, Slater,* and *Reynolds* between 1862 and 1871, introduced and recommended cells in which *cast-iron* formed one of the elements. When *this* is used instead of the platinum or carbon, these cells may and do sometimes give a powerful current, especially when dilute hydrochloric acid is employed in place of sulphuric acid as the excitant along with the zinc. But the nitric acid employed as the depolarizer frequently becomes suddenly heated, attacks the iron violently, giving out at the same time copious fumes of nitrous acid gas, and causing the nitric acid to boil over, to the destruction of everything in the vicinity.

Perhaps the best of the double-fluid cast-iron cells is the one described in the next section, which is due to

Turton. As it contains no nitric acid, it has at least the merit of being free from the boiling-over trouble.

§ 179. *Turton's* cast-iron cell (1867) is one in which cast-iron takes the place of the carbon. It may be used either as a single or double-fluid cell. In the former case the exciting solution, in which the amalgamated zinc stands, consists in $3\frac{1}{2}$ parts of a saturated solution of common salt mixed with 1 part of strong sulphuric acid. The E.M.F. is low, owing to the counter E.M.F. of the cast-iron; but the current is large. As a double-fluid cell, the same exciter may be used in a porous cell along with the zinc, while a saturated solution of perchloride of iron between the porous post and the cast-iron acts as the depolarizer.

§ 180. In 1890 the *Weymersch* cell was put prominently before the public, and much money was lost by those who interested themselves in it. Like many other highly-vaunted lighting cells, it was nothing but a double-fluid bichromate. An amalgamated zinc plate stood in a dilute solution of sulphuric acid (1 part acid and 18 of water). The depolarizer was a secret mixture of sodium nitrate, chromic and sulphuric acids. The elements were arranged in a rectangular trough of ebonite or wood, $9\frac{1}{2}'' \times 11'' \times 13''$, divided into six compartments by ebonite partitions. All these are connected together by short tubes to one common exit pipe, to permit of drawing off all the liquid in the cells at one operation when required. In each compartment stood a quart-size porous pot, fitted with ebonite syphon likewise connected to a common draining-pipe. In the porous cell is placed the carbon plate or rod; in the outer compartment the amalgamated zinc. Two six-gallon stoneware jars, the one 'filled with the exciting solution and the other with the depolarizer, were placed on a shelf above the cells, and the liquids allowed to flow, through indiarubber tubes, to their respective compartments, until the cells were filled up to a marked level. The cells were then ready for action. Six cells, as above, were stated to give 13 volts, with a working current of about 15 ampères for seven hours. When the solution gave signs of exhaustion they were drawn off by means of the two tubes, and fresh fluid allowed to run in from the stone jars.

§ 181. *Mons. Rouillon* (1867), having observed that a mixture of nitric and hydrochloric acids, in certain proportions, though capable of dissolving gold and platinum freely, has little or no action on metallic silver, constructed a modified form of the Grove or Bunsen cell, by using silver as the negative element, in the place of the platinum or graphite. The exciting solution, acting on the zinc, is as usual a dilute sulphuric acid (1 acid, 12 water, by weight), while the depolarizer consists in an "aqua regia" containing equal parts by weight of strong nitric and hydrochloric acids. Several months' trial tended to prove that the superficial coating of silver chloride formed on the silver plates protects them from the action of the nitric acid, without appreciably lowering their conductivity: while the close texture of the silver effectually prevents the acid creeping up to the terminals. Mons. Rouillon considered this battery to be superior to the original Bunsen.

§ 182. In *Cliff's* modification of the Bunsen cell, and also of the Grove cell, a perforated tray is fitted over the top of the cell, which tray is filled with lumps of quick-lime. The purpose that is served by the quicklime is the absorption of the objectionable nitrous fumes (see § 154).

§ 183. *Prevost* modified the Bunsen by placing a conical zinc cylinder in the outer containing vessel along with the dilute sulphuric acid, suppressing the porous pot and substituting a carbon pot, which formed at once the negative element and the porous division, this latter containing a patented substitute for nitric acid, consisting essentially of a mixture of nitrate of soda and sulphuric acid.

§ 184. *Mr. Agapis* recommended in 1880 a modification in which the zinc stands in a solution containing 15 per cent. of potassium cyanide, the carbon being immersed as usual in nitric acid. He claims for this that the zinc requires no amalgamation, that while the E.M.F. is very high, and the current constant, no nitrous fumes are given off. This may be true, but we are afraid the remedy is worse than the disease, since hydrocyanic acid (prussic acid) is evolved, and this is far more dangerous than the nitrous fumes.

§ 185. *Blackwell* of Liverpool modified the shape of the Bunsen cell by lining the sides of a square outer containing vessel with sheets of zinc, the carbon standing in a central porous pot as usual (1881).

§ 186. *Mr. Willrants* in 1882 invented a cell in which the zinc stands in a concentrated solution of ammonium chloride, and the carbon in a saturate solution of sulphate of iron contained in a porous cell; the upper extremity of the porous cell and also of the carbon should be greased, and the strength of the sulphate of iron solution should be maintained by keeping a few crystals of sulphate of iron in a perforated tray at the top of the porous cell. The E.M.F. of this cell is about 1·8 volts, the internal resistance is rather high, being about 1 ohm in the pint size. On the other hand it will run for nearly six hours before it becomes polarized, recovering itself after about half-hour's rest. The following is the reaction which takes place :—

$$4\,Zn + 8\,NH_4\,Cl = 4\,Zn\,Cl_2 + 8\,NH_3 + H_8.$$

This hydrogen seizes on the portion of the oxygen contained in the ferric sulphate, $Fe_2\,3\,SO_4$ (produced by the action of the air on the ferrous sulphate), reducing it again to the ferrous state :—

$$H_8 + Fe_2\,3\,SO_4 = 4\,H_2\,O + 2\,Fe\,SO_4.$$

This latter again absorbing oxygen from the atmosphere, after a little exposure thereto.

§ 187. *M. Gerardin* devised in 1866 a battery, using iron turnings, into which a plate of iron is thrust, and immersed in water only, for the usual zinc electrode, and having a carbon electrode in a porous pot filled with a solution of perchloride of iron, to which "aqua regia" has been added. The carbon electrode is said to have been made by crushing retort graphite and agglomerating with paraffin. It was said that a great amount of galvanic power was obtainable at a small cost. E.M.F. about 1 volt. Resistance not stated. The reaction is as follows :—

$$Fe_2 + 3\,H_2\,O = Fe_2\,O_3 + 3\,H_2.$$

This hydrogen seizes on a portion of the oxygen absorbed by the ferric chloride from the air: a little rest re-converts the ferrous salt again into ferric.

§ 188. Another form is due to *Callan*. It consists in a round outer cast-iron cell, with flanged base for steadiness, and a lug cast on for connections. There is a porous jar with zinc cylinder inside, acted on by sulphuric acid 1 part to 8 parts of water. Solution in iron compartment, 2 parts sulphuric acid, 2 parts nitric acid, and 1 part water. Using pots 7″ high and 4″ diameter, with large porous cell containing a rod of well-amalgamated zinc having a wire cast into it, connection being made by the wire from zinc dipping into mercury contained in a hole drilled in the lug on cast-iron cell; 40 cells in series, instantly melted copper wire $\frac{1}{8}$″ diameter, producing green smoke or vapour. Platinum wire nearly $\frac{1}{16}$″ thick ran into drops, but was not vapourized.

§ 189. In *Slater's* soft iron battery we have an iron plate or cylinder immersed in a saturated solution of potassium nitrate, the carbon rod or plate standing in a porous cell filled with nitrous acid of the specific gravity of 1·6, which should reach about 1″ higher than the fluid on the outside. By this means the percolation of the acid through the porous cell keeps the solution in the outer jar sufficiently acid. Another depolarizer recommended by Slater consists in a mixture of equal parts by measure of strong nitric acid of the specific gravity of 1·6, and of hydrochloric acid of the specific gravity of 1·2. If used for telegraphic work, a solution of nitric of potash or of soda can be used in both compartments. To obtain an increase in E.M.F. without employing more cells the addition of $\frac{1}{10}$ of acetic acid will be found sufficient.

§ 190. In 1882 *Rousse* introduced his ferro-manganese cell with a view of obtaining as well as electricity, salable by-products. Instead of zinc he used ferro-manganese of 80 per cent. in dilute sulphuric acid, and carbon in concentrated nitric. If intended for working in ordinary rooms a solution of permanganate of potash can be used instead of the nitric acid. The E.M.F. in either case is about 2·2 volts. To recover salable products he proceeds as follows :—The salts produced in working the cell are sulphate and nitrates of manganese and potassium. To remove the sulphuric acid, Mr. Rousse treats the liquid with lead nitrate (which is produced in his "lead" cell). The lead sulphate so formed is boiled with potassium carbonate, which produces *Ceruse* (white lead). The soluble salts separated by decantation only consist of nitrates of potassium and manganese. The manganese is procured as carbonate by adding carbonate of potash, and is then washed and calcined, which forms sesquioxide of manganese. This is heated with potassium hydroxide and nitrate to reproduce permanganate of potash.

§ 191. In the *Ross* cell (1883) has been introduced a novel idea of *nickelling* the zinc instead of amalgamating it, to prevent local action. The constituents are nickelled zinc and carbon in a special solution made of 1 volume of hyponitric acid, 3 volumes of sulphuric or $4\frac{1}{2}$ volumes of hydrochloric acid, and 4 volumes of water, in carbon compartment. The zinc stands in a solution consisting of 1 part of sulphuric acid, or $1\frac{1}{2}$ parts of hydrochloric acid in 100 parts of water. It is said to be not so powerful as the Bunsen, and to give off chlorine, if hydrochloric acid be used in either of the solutions.

§ 192. In 1884 a patent was taken out by *Messrs. Prizibram, Scholz and Wenzel*, of Vienna, for a galvanic element composed to two electrodes, which may either both be of metal, or one of them may be of a mineral substance, and three conducting liquids, these parts being disposed in a vessel divided by means of two diaphragms, or partitions, into three compartments. The positive electrode, which is made either of platinum, carbon, or silver, is placed in a compartment of the element containing pure nitric acid of from 30° to 40° Beaumé. The negative electrode is placed in a second compartment of the element, containing a solution of one part by weight of common salt in about 20 parts by weight of water, or filled with some alkaline liquid. The third compartment of the element separated from the other two by a porous diaphragm made of clay, or other suitable porous substance, contains concentrated sulphuric acid of from 50° to 70° Beaumé. The object of the inventors was to find an improved constant cell of equal power to the Bunsen, but as they failed to complete the patent, it may be assumed that the results they obtained were not so satisfactory as they expected.

§ 193. *F. W. Maron's* regenerative ferric chloride cell is practically a Bunsen in which a solution of 1 pound of ferric chloride in 1 gallon of water substitutes the nitric acid. The E.M.F. is about 1·5 volts; it gives off no fumes; and when the depolarizer is exhausted it may be restored to activity by passing a current of chlorine gas through it.

§ 194. *M. Baron's* cell. In this the zinc is placed in a porous cell containing a solution of common salt or ammonium chloride, having a strength of 15 grammes of salt to 10 litres of water. A carbon cylinder stands outside this in a special mixture made as follows: 1 part of powdered wood charcoal, 5 parts of water, $1\frac{1}{2}$ parts of sulphuric acid, and $\frac{1}{2}$ part of zinc. The addition of this latter causes bubbling. To this is added little by little a $\frac{1}{4}$ part of pure litharge or minium. The mixture is allowed to bubble for three hours, after which it is filtered from the charcoal; $1\frac{1}{2}$ parts of strong nitric acid are now added, when the solution is ready for use. This cell gives about 2 volts, and 6 cells are capable of running an 8-volt lamp for 12 days of 24 hours without interruption. Fumes, however, are given off. If it be desired to do away with these fumes, tartaric acid in the proportion of $1\frac{1}{2}$ parts must be used instead of the nitric acid; but in this case a $\frac{1}{4}$ part of sal ammoniac must be added to the solution.

§ 195. Of the mercury sulphate type, there is only one which has received any extended application as a double-fluid cell. The same may also be said of lead sulphate cells. We will therefore describe both these together as they are very similar.

§ 196. *Marié Davy* cell. Double-fluid. Constituents:—Amalgamated zinc in acidulated water, and carbon in mercury sulphate paste. The arrangement is, the zinc is placed in a solution of salt or of acidulated water standing in an outer vessel; in this solution is placed a porous pot which contains a carbon plate surrounded by a paste of mercury sulphate and water. This latter is made by agitating the mercury sulphate with about 3 times its volume of water, the clear liquor is then poured into the porous cell, so as to allow it to permeate the pores. The remainder of the paste is then used to fill in the interstices between the carbon plate and the sides of the porous cell. The E.M.F. of this cell is 1·52 volts. No action takes place on open circuit; if used on a closed circuit, with a low resistance, depolarization does not take place quickly enough to enable it to supply a large current; but on a high-resistance circuit it is very constant. In a special trial made with 38 of these cells against 60 Daniell cells of the same size, it was found that the Marié Davy cells could be used for 5 months 23 days consecutively, while the Daniell cells only lasted 2 months and 23 days. In using these cells care must be taken that the level of the liquid be kept the same in all, otherwise that cell in which the level is lowest will be polarized and exhausted. The action may be expressed as follows:—

$$Zn + H_2 SO_4 = Zn SO_4 + H_2.$$

Then in the mercury compartment:—

$$H_2 + Hg SO_4 = H_2 SO_4 + Hg.$$

The mercury thus liberated can be collected and re-dissolved in sulphuric acid to form fresh sulphate. Some of the mercury sulphate passes through the porous pot, and helps to keep the zinc amalgamated.

§ 197. *Messrs. King Mendham and Co.*, of Bristol, about 1885 used to supply a lead sulphate cell consisting of a zinc rod or plate standing in a porous cell containing dilute sulphuric acid of the strength of 1 part of acid to 12 of water by weight; in the outer vessel was placed a cylinder of sheet-copper immersed in a saturated solution of lead sulphate. As lead sulphate is practically insoluble in pure water, we presume that sodium thiosulphate (hyposulphite of soda) was employed as the solvent. The E.M.F. of this cell is given as 0·55 volt.

We now come to three cells using the haloid elements (which are greedy of hydrogen) as depolarizers. These are :—

§ 198. *Upward's* chlorine cell. Constituents: zinc and carbon, in zinc chloride, using chlorine gas as a depolarizer. Arrangement :—An outer oblong glazed earthenware jar, in which stands a suitable porous cell. The outer jar contains two plates of carbon, packed with coke or retort scurf. There is an outlet with a tap in the bottom of the jar, to drain off any solution percolating through from the porous cell which keeps the carbon moist. The outlet pipe should always *dip* into the liquid drained off, to form a water seal. A constant stream of chlorine gas is forced into the outer cell at the bottom, and leaves it for the next cell at the top. The carbon cells are hermetically sealed. The porous cell contains a zinc rod in water. To start the battery, a partial vacuum is formed in the carbon compartments, to draw in a supply of chlorine from the holders. This acts on the zinc, forming zinc chloride, which percolates through the porous cells, moistens the carbon granules, and drains off into a trough. Air is unavoidably drawn into the battery and accumulates in the last cell. It is disposed of by an ingenious device consisting of an electro-magnet actuating an aspirator that draws off all the air. As soon as all the air has been removed, the E.M.F. increases and cuts off the current from the magnet, thus stopping the aspirator. The E.M.F. is 2·1 volts. The resistance of a cell $11\frac{1}{2}'' \times 5\frac{1}{2}'' \times 12\frac{1}{2}''$ is only 0·2 ohm, hence the current is 10

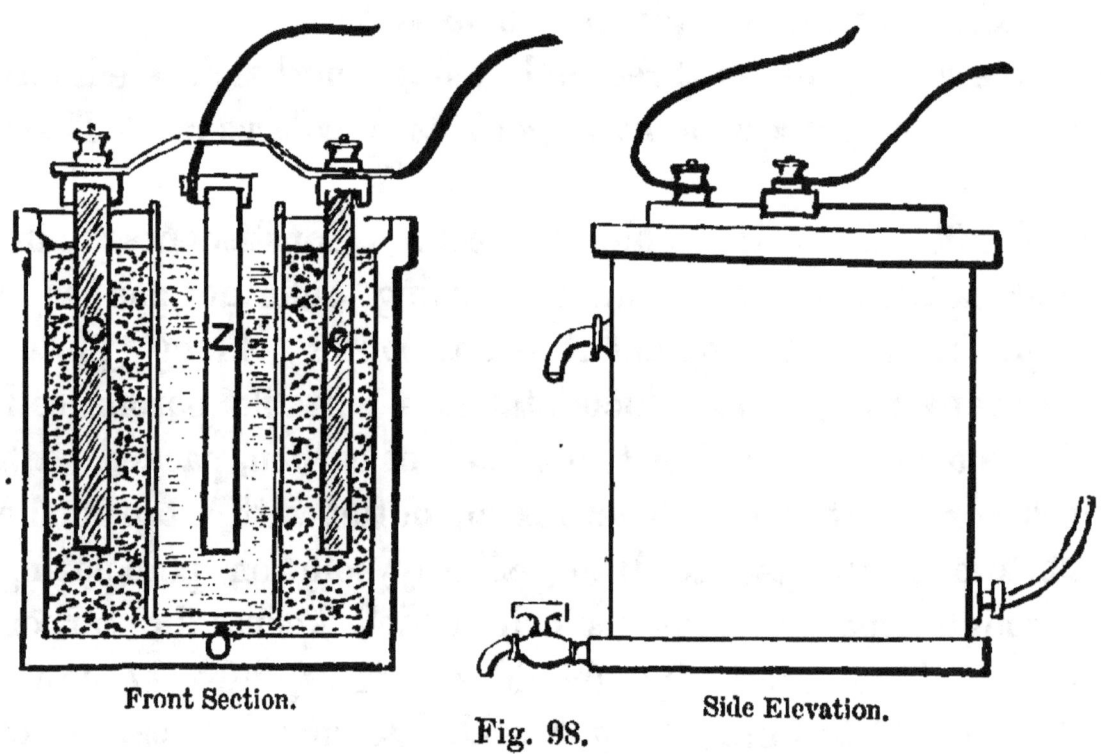

Front Section.   Side Elevation.

Fig. 98.

ampères, constant as long as the chlorine is supplied. The water in the zinc cells wants replenishing once a week. The gas was made by heating a mixture of brine and manganese peroxide in proportion of 3 parts of manganese dioxide to 4 parts of common salt and sulphuric acid in an earthenware jar fixed in a sand-bath. Fig. 98 gives a general idea of the arrangements of the Upward cell.

§ 199. In 1882 Professor *Exner* instituted several

experiments with Bunsen cells, using bromine and iodine instead of nitric acid as the depolarizer. He employed carbon always as the negative plate, and varied the positive metal. The annexed table shows the results he obtained with the different metals, in conjunction with bromine, or iodine in solution in water. There was no perceptible polarization, and the current was steady and constant as long as there was any depolarizer left.

Positive Element.	Negative Element.	Depolarizer.	Observed E.M.F.	Calculated [1] E.M.F.
Magnesium	Carbon	Bromine	2·36	—
Aluminium	,,	,,	1·60	1·61
Zinc	,,	,,	1·52	1·52
Lead	,,	,,	1·29	1·29
Silver	,,	,,	0·91	0·91
Copper	,,	,,	0·51	0·65
Platinum	,,	,,	0·04	—
Magnesium	,,	Iodine	1·57	—
Aluminium	,,	,,	0·77	0·93
Zinc	,,	,,	0·96	0·98
Mercury	,,	,,	0·55	0·68
Silver	,,	,,	0·56	0·55
Platinum	,,	,,	0·313	—

§ 200. The *Bromine* cell, devised by the author in 1900, is prepared by putting a porous cell in the centre of a larger glazed earthen or glass containing vessel, size about 6″ × 4″, and packing this round with granular carbon in which is inserted a carbon plate or cylinder fitted with a terminal. About an ounce of bromine is poured over the carbon, after which the outer vessel is filled to within an inch of the top with dilute hydrochloric acid. A well-amalgamated zinc rod stands in the porous cell. The excitant is a 5 per cent. solution of chloride of zinc. There

[1] See § 13 for the data on which these calculations are based.

is no local action, and no action on the open circuit. When the circuit is closed the E.M.F. is nearly 3 volts. Fig. 99 illustrates this cell.

§ 201. The *Iodine* cell, as constructed by the author, is very similar to the above. But in order to keep up a fair supply of iodine, the following mode of putting up the cell

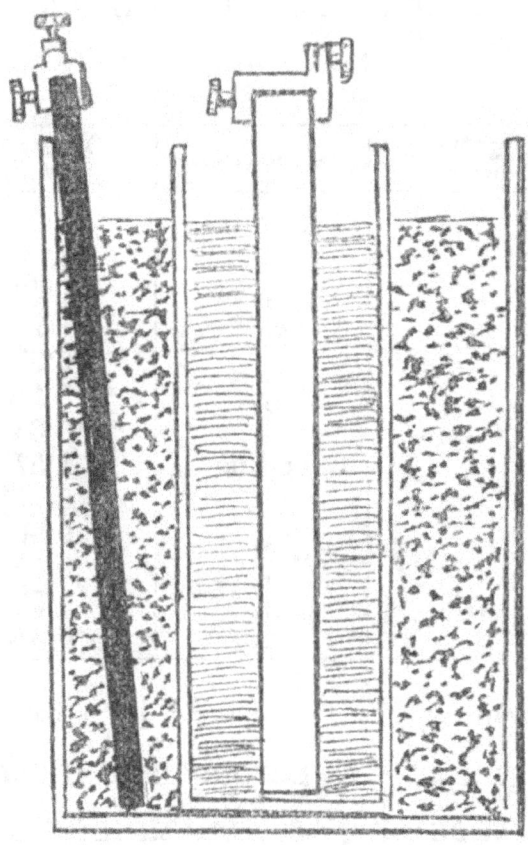

Fig. 99.

is recommended. A circular disc of carbon nearly as large as the containing cell is fitted with a central carbon stem, reaching about 1″ over the level of the cell. This is furnished with a terminal. About an ounce of iodine in small crystals is scattered over the surface of the carbon disc, and over this is poured about 1 oz. of a saturated

solution of potassic iodide.   To prevent too rapid mixing
of the fluids, a thin layer of silver sand is sifted over the
iodine.   The vessel is then filled, to within 1 inch of the
brim, with either dilute hydrochloric acid (1 part acid to
12 of water) or with a 5 per cent. solution of chloride of
zinc.   A stout zinc cylinder furnished with a terminal, and
three equidistant lugs, to support it on the edge of the cell,
is now immersed in the cell.   This zinc cylinder should
extend downwards to within an inch of the carbon disc.
E.M.F. 2 volts.   Very constant until all the iodine is exhausted.   An excellent cell for long steady work, in which
a fairly high E.M.F., but not a very heavy current, is
required.   The arrangement of this cell is shown in section
at Fig. 100.

§ 202. The *Ettore* chlorine cell is very similar to the
preceding, except in arrangement.   The elements are zinc
and carbon.   The exciting fluid is a nearly saturated
solution of sal ammoniac in water; the depolarizer is
chlorine in a gaseous form, produced by allowing hydrochloric acid to fall drop by drop on chlorohypochlorite of
calcium (the "chloride of lime" of commerce) contained
in a closed vessel connected to the battery-box by means
of a tube.   By means of an exit-tube at the top of the
closed battery-box, any free chlorine is led into a receptacle
containing lime, which absorbs it.   The battery-box entering the cells is hermetically sealed.   The carbons are
made in the shape of a porous pot, and they contain the
excitant and the zinc, which should be amalgamated.   The
chlorine gas acts from the outside of these porous carbon
cells.   The E.M.F. is high, and the internal resistance very

low.   Each cell gives 2·5 volts, with a resistance of about 0·1 ohm.

The following cells, though of some interest theoretically, have not as yet received any extended practical application.   The " heat " cells, the " light " cells, and the " gas " cells may possibly find fields for usefulness.

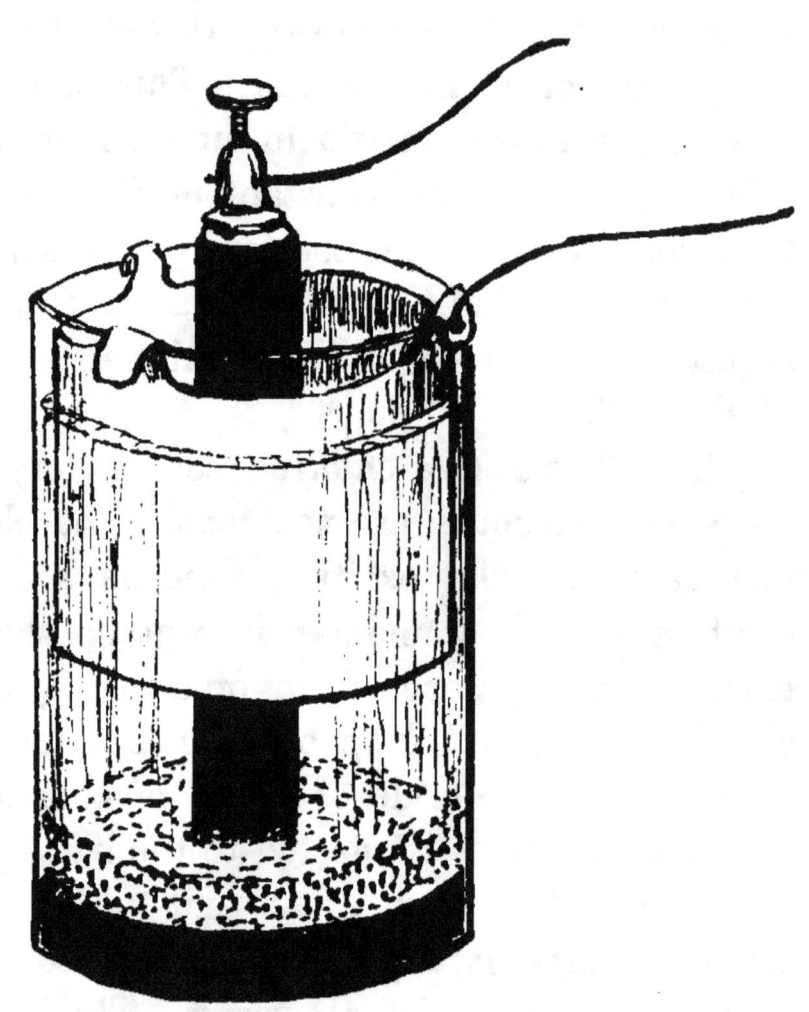

Fig. 100.

§ 203.   *G. G. André's* cell.   Constituents:—Lead packed in carbon and lead peroxide.   Zinc, iron, or lead, outside porous partition, and dilute sulphuric acid.   Arrangement:—The lead is wound as sheet or wire on a wooden core, having a bottom flange, and is packed round with highly-burnt coke and lead peroxide. The whole is placed in a porous envelope, which may consist of two thicknesses of canvas, nitrated to make it acid-proof if desired. The idea is that the negative element (which is the positive pole) may be re-vivified at a charging-station by the re-oxidation of the lead peroxide when reduced; also that the solution may have the zinc which has been dissolved re-precipitated. This cell is therefore a form of accumulator, not very dissimilar to the " Kingsland " cell. When excited by dilute sulphuric acid, the E.M.F. is high, nearly 3 volts, and the internal resistance being small, a large current can be taken off.

§ 204. *Fournier's* cell. Constituents :—Copper and a composite plate in dilute sulphuric acid. The copper is attacked by the sulphuric acid on completing the circuit, liberating hydrogen and forming copper sulphate. The hydrogen goes to the composite plate, which is positive outside the liquid, and there is oxidized by the lead peroxide contained in it. The plate is formed of lead protoxide, lead peroxide, and glycerine. The lead protoxide is used because peroxide alone does not "set" well with glycerine. The peroxide conducts much better. Constant, until all the oxide is reduced, when a current passed in the reverse direction will restore it to the original condition. Very similar to the last, except that the E.M.F. is much less, as copper is lower in the scale than is zinc.

§ 205. *Messrs. Alfred Birn and F. Flasslachers* (1886) contrived a constant cell in which the zinc stood in soda-lye and the carbon in aqua regia, this latter either alone, or mixed with ferric chloride. If chromic acid be used with the nitric acid in the aqua regia instead of hydrochloric acid, platinum or aluminium may be used as the negative element instead of carbon. The arrangement is :—Zinc in the outer cell with soda-lye of varying concentration. Carbon, platinum, or aluminium in the porous cell with aqua regia or the above-named mixture. E.M.F. 2·5 volts, resistance very low. The reaction is as follows :—The zinc is oxidized, and the resulting oxide is dissolved by the soda-lye. Hydrogen is given off during this reaction, and seizes upon the chlorine of the aqua regia, forming hydrochloric acid. Hence it is necessary to add nitric acid from time to time. If chromic acid be used, the hydrogen seizes upon the oxygen thereof, giving rise to the production of water, and chromium sesquioxide, which either enters into combination with the nitric acid to form chromous nitrate, or, if nitric acid be in excess, is again oxidized to the chromic state.

§ 206. To the alkaline excitant, double-fluid class belongs the "*Walker Wilkins*" (1894) cell. In this, as shown at Fig. 101, we have an earthenware tray U, fitted with a draining-pipe D. O is an outer stoneware vessel, having no bottom, but incurved edges, which rests on the tray U. A cylinder of perforated nickel N rests also on the ledge, at about half-way between the jar O and a porous pot P, which rests on the extreme edge of the incurved piece. This porous cell P is also furnished with a little indiarubber draining-pipe E, fitted with a spring cock. The space between the outer jar O and the nickel cylinder is packed with coarse carbon C, and that between the nickel and the porous cell with finely-powdered graphite G. In the porous cell stands a zinc cylinder Z, to which is attached one terminal $T-$, the exciting solution being a solution of caustic potash in water, of the strength of 12 oz. to the quart. This is shown at K in our illustration. Another terminal $T+$ is attached to the nickel

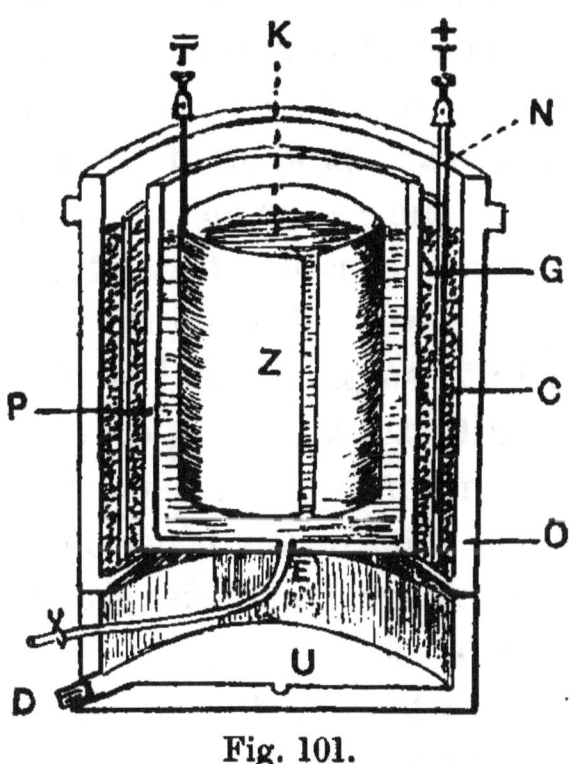

Fig. 101.

cylinder. It is advisable to float a layer of heavy petroleum oil over the caustic potash solution, to retard the absorption of carbonic acid gas from the atmosphere. E.M.F. 1·3 volts, internal resistance in quart-size cell 0·6 of an ohm. Said to be absolutely self-recuperative. Prof. Varley states that he had one such cell running for five months consecutively, without depreciation. This cell is related to the Lalande Chaperon, § 89, *q. v.*

§ 207. *Highton*, of Putney, in 1871, published an account of several batteries of his invention, of the most novel of which we here present a brief description. The "Cheap" cell:—The negative element consists in a plate of carbon standing in a porous cell, in which is also packed pieces of light cinders, about the size of a hazel nut. The excitant is strong nitric acid, or a mixture of oil of vitriol and potassium nitrate in equal proportions. For the positive element, a carbon plate stands in a solution of any soluble sulphide, by preference (for cheapness) the sulphide of calcium, which is a waste product of the soda works. The "Steam" battery is one in which an outer vessel of any non-metallic substance capable of standing heat is fitted with a central porous cell. Around this is packed coke in large lumps. Dilute sulphuric acid is placed in this, not sufficient in quantity to cover the coke. In the porous cell is placed a zinc or iron element, with a saturated solution of common salt (or any other neutral salt) as the excitant. The whole arrangement is placed over a Bunsen burner. This causes the evolution of steam from the surface of the coke, which in the words of the inventor "sweeps away the hydrogen," so that the battery becomes perfectly constant, and the heat causes it to be twice or three times as energetic as the ordinary form.

§ 208. In *Highton's* zinc-carbon cell we have yet another form of alkaline excitant battery. The carbon element is placed in a porous pot and packed round with milk of sulphur (sulphur precipitated from an alkaline solution by means of an acid), with which is mixed granular carbon and manganese dioxide, moistened with dilute sulphuric acid.   The zinc is placed in an outer compartment, in which is placed a strong solution of caustic potash or caustic soda.   The inventor claims for this cell an E.M.F. 50 per cent. greater than the Grove cell, viz. 3 volts, but the resistance is high.   The presence of *sulphur* in a cell *may* facilitate the absorption of the polarizing hydrogen, for it is a well-known fact that sulphur readily combines with nascent hydrogen, with the production of that vile-smelling compound, sulphuretted hydrogen (hydric sulphide); but what militates against the supposition that any such reaction takes place in this case, is the fact that Highton claims that no noxious fumes are given off.   First described by Highton of Putney in 1872.

§ 209. *Rapieff's* cell (1883).   Constituents:—Zinc and lead in a porous pot packed with peroxide or oxide of lead, barium, manganese, silver, iron, copper, separately or together, and with or without lead clippings, shot, or carbon granules, in a special solution.   Arrangement:— The lead may be in any form, packed as above, and the cells sealed except for holes to let out any gas produced during charging and discharging.   The zinc may be the material of the outer cell, if desired.   The solution is one of hydroxides of sodium, potassium, or other alkaline metal, or of sulphuric, hydrochloric, oxalic, chromic or other acids, or particularly when peroxides or oxides of lead, iron, silver, or copper are used, a solution of chloride, chlorate, permanganate, sulphate, nitrate, or carbonate of ammonium, sodium, potassium, or manganesium.   Or as a two-fluid cell, with carbon, lead or other metal, a 15 to 30 per cent. solution of sulphuric, hydrochloric, or chromic acids, in absence of manganese oxides. The exciter in zinc cell is a solution of salts of ammonium, potassium, sodium, magnesium, or any other alkaline metal, or even zinc sulphate.

§ 210. *S. H. Emmen's* cell (1886). Constituents :—Zinc and peroxidized lead plate in dilute sulphuric acid. These peroxidized lead plates are re-vivified by being used as anodes in a copper sulphate bath.

§ 211. *Mr. Lyte* (1885) patented a cell having a lead box full of peroxide of lead at bottom as + pole, and a coil of lead wire at the top supported by the cover as the − pole.

§ 212. *Jablochkoff's* "anti-accumulator" or three-electrode cell (1885). Constituents :—Lead, another more oxidizable metal, and carbon, in a moist mass of sawdust, etc. Arrangement :—A tank of lead, or one lined with lead, 1 decimetre square, and 25 millimetres deep. This tank contains pieces of sodium, sodium amalgam, iron turnings, or any oxidizable metal in an otherwise useless state, and is then filled nearly to the top with any spongy matter, such as cloth, or wood sawdust, saturated with water (if sodium be used), or a solution of sodium chloride (common salt), or preferably calcium chloride, as it is very deliquescent. If sodium and water are used, sodium hydroxide is formed, which attracts moisture, so keeping the cell moist. On the top of the spongy mass is placed a range of porous carbon tubes (a perforated carbon tank may be used with advantage). When the circuit is open, local currents are produced between the oxidizable metal and the tank, which becomes polarized, and its potential rises until it equals that of the metal. To draw off a current, wires are connected to the tank and the top layer of carbon. E.M.F. with sodium amalgam 2 volts, with zinc 1·6 volts, with iron 1·1 volts. May be left unused for long periods, and when refreshment is necessary it is sufficient to immerse the battery (usually of ten cells in series) in water, remove it, and then immerse in calcium chloride solution. Good for bells and telegraphs, also for light or power if renewed every 48 hours or less. Said to give 1 H.P. for 5 centimes—about ½d. Resistance ¼ to ½ ohm; weight, 200 to 250 grammes.

§ 213. In the United States Treasury, for the portable lighting of safes, etc., in 1895, was used a modification of the silver chloride cell, known as the *Capo Farad*. It consists in a zinc tube 2¾″ long, 1″ in diameter, fitted with a stopper of indiarubber, which is perforated to allow the passage of a stout silver wire. On this silver wire is cast a cylindrical mass of silver chloride, surrounded by a casing of any textile material soaked in zinc chloride solution. The cased silver wire may be supported if needful by means of a couple of ebonite washers, and the space between the casing and the inside of zinc tube filled in with a solution of zinc chloride. E.M.F. 1·5 volts. It is practically a "dry" cell. Four such cells in series weigh 5 ounces, and will light a six-volt lamp, at intervals, for a considerable period. Will work in any position.

§ 214. *T. Slater*, of iron cell repute, also devised a "nickel" cell (see Fig. 102). The nickel stands in a solution of nickel sulphate, or of ammonia sulphate of nickel. The carbon is placed in a porous pot containing

1 to 12 solution of sulphuric or hydrochloric acid. Sometimes used as a circulating cell, like Ponci's, § 158, *q. v.*, in which case it is better that the nickel should occupy the central position in the porous pot, and blocks or plates of carbon (connected together) take the outside place. No details are given as to E.M.F. or resistance.

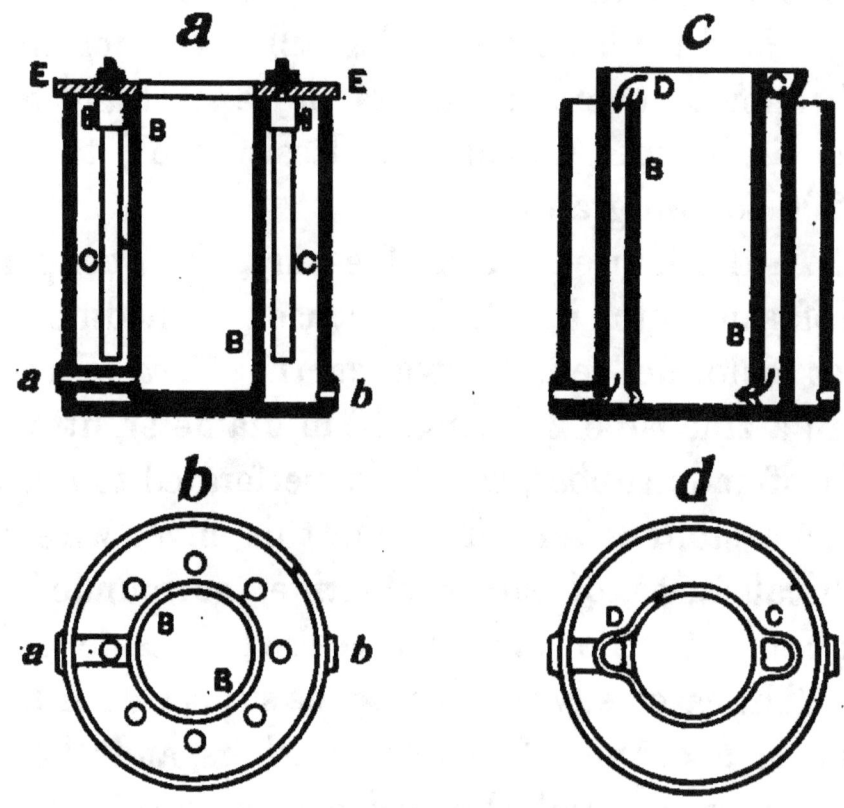

Fig. 102.

§ 215. In *Wirt's* constant cell (1882) we have a peculiar arrangement of three square bottles, fitted with stoppers. These are placed in a wooden frame, only just too long for them to fit tightly. In the upper part of the adjacent sides of bottles No. 1 and No. 2 is drilled a hole $\frac{3}{8}''$ in diameter. Similar holes are drilled at the bottom part of adjoining sides of No. 2 and No. 3. Between each of these are placed indiarubber washers, which make water-tight joints, and which also keep in position diaphragms of goldbeater's skin. A copper electrode is placed at the bottom of No. 1, in a dilute solution of zinc sulphate, also a crystal of copper sulphate, the size of a pea. In bottle No. 2 is a solution of zinc sulphate, with a piece of zinc to absorb any copper sulphate that may percolate. In bottle 3 is the zinc electrode, suspended from the stopper. Said to be absolutely constant.

§ 216. *Higgin's* cell (1882) is a modified Smee, with a mixed depolarizer and an air-blowing device. Constituents:—Tin and carbon plates in a solution consisting of dilute sulphuric or hydrochloric acid and persulphocyanide of iron. The iron solution is made thus:—Dissolve ferric sulphate or chloride in water, and add thereto an alkaline or earthy sulphocyanide solution in chemical proportion to the iron. Arrangement:—A bar of iron rests by its ends on the ends of a rectangular cell, and is secured to a tin plate. One way of doing this is to insert bits of copper wire in holes in the tin plate, and fix them by running in melted tin. The ends of these wires have screw-threads cut in them, and they pass through the iron bar, and are secured by nuts.

Two pieces of gas carbon are secured one on each side of the tin plate, insulated from it by strips of any insulating material such as vulcanite; the whole is fastened together by a clamp and terminal combined, in the same way as a Smee's cell is. At one end of the cell, close to the bottom, is a piece of quill tubing midway between the sides, and curled over at the top, which acts either for blowing air through the solution, or for draining the solution away. Said to be powerful.

§ 217. In 1882 Messrs. J. and A. J. Higgin, of Manchester, patented several batteries, in which *tin* formed the electro-positive element to the exclusion of zinc. A carbon plate or rod is fitted in a porous cell, and packed with pyrolusite and crushed carbon, the cell being then filled up with dilute sulphuric acid, 1 to 5 of water. A tin rod, plate, or cylinder stands outside the porous cell in the containing vessel. The excitant is dilute sulphuric acid. Or, the pyrolusite and carbon granules may be replaced by a solution of persulphocyanide of iron. During the action of the cell, the tin combines with the acid to form stannous sulphate, which can be converted into stannic oxide by treatment with a sufficiency of manganese dioxide, and subsequently with an excess of water, which precipitates stannic acid, which is salable. Even the spent depolarizing solution can be utilized, since the addition of a copper salt to the filtered solution (previously deoxidized by means of sulphurous acid or other deoxidant) causes the precipitation of disulphocyanide of copper, which is used in calico-printing. This is yet another attempt to solve the problem of getting work out of a battery for nothing. But the purification of the waste products is too costly to render the result satisfactory.

§ 218. *Martyn Roberts* "Tin" cell. With the view of producing a marketable by-product during the action of the battery, it was proposed in 1870 to use *tin* instead of zinc as the positive element in a cell, carbon being the negative. In this case no porous cell need be used, both elements standing in a dilute solution of nitric acid, sp. g. 1·24. No deposit is formed either on the tin or the carbon, the result of the reaction being metastannic hydrate, which is precipitated and deposited at the bottom of the vessel. The E.M.F. is stated to be "as strong as Grove's" (2 volts). The advantages claimed for this cell are—1st. The metal dissolved is precipitated, thus not altering the exciting fluid (?). 2nd. The precipitate thus produced is a salable article. This may be true, but it is questionable whether a much larger current could not be got much more cheaply from zinc and carbon than from the more expensive metal *tin*, even with the deduction of the value of the "salable product."

§ 219. *Messrs. A. F. St. George* and *C. R. Bonne* patented in 1889 a very novel form of cell, in which a central carbon pot containing nitric acid is placed within a larger vessel of metallic tin (not tinned iron). All round and under the carbon pot is packed a paste of metastannic hydrate, $H_{20} Sn_5 O_{20}$. This is prepared by the action of ordinary commercial nitric acid on tinfoil. A violent chemical reaction takes place, with the production of metastannic hydrate in the form of a damp white powder. As the nitric acid in the carbon cylinder would attack any ordinary metal, the metal collar that encircles the top of the carbon, and which carries the terminal, is faced with platinum foil. The other terminal is connected to the tin cylinder. The damp paste acts with great chemical affinity on the tin cylinder, and when the circuit is closed the galvanic action which takes place reduces the metastannic hydrate to the condition of metallic tin, or of a lower oxide of tin. The nitric acid oozing out of the sides and bottom of the carbon vessel re-converts the metallic tin, or the tin oxide thus produced, into meta-stannic hydrate, ready to undergo a similar cycle of changes. It is therefore necessary to keep up the supply of nitric acid. A very neat way of arriving at this end is to stand the entire cell in a double-jacketed outer vessel

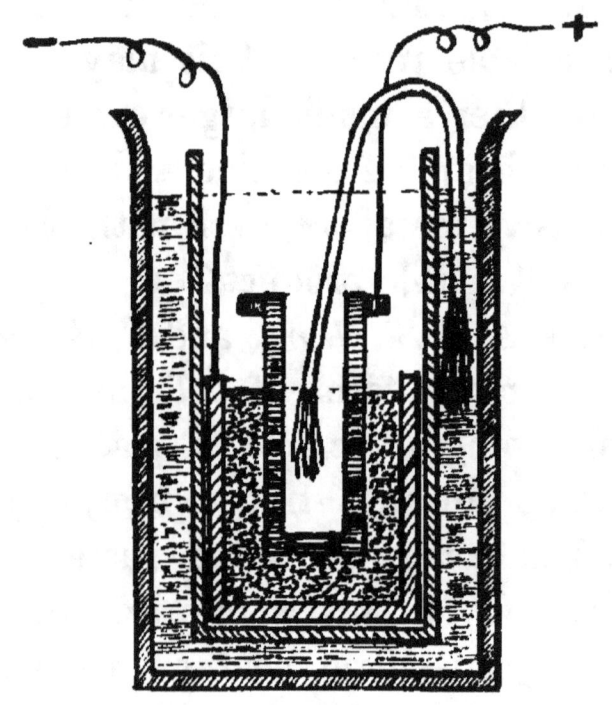

Fig. 103.

(see Fig. 103), this latter containing a supply of nitric acid. A glass syphon, with an asbestos wick at each end, conveys little by little the nitric acid from the outer vessel into the interior of the perforated carbon pot, whence it slowly percolates through to the metastannic paste. The E.M.F. is said to be high, and the current practically constant as long as the supply of nitric acid lasts.

§ 220. In *Faure's* "regenerating cell" we have an entirely new departure. The constituents are carbon and iron, in a solution of common salt, into which is fed a stream of carbon dioxide. A strong wooden box or trough 25 metres long (about 27 yards), 2 metres wide (6 feet 6 inches), and 1·2 metres in height (4 feet), contains 100 double electrodes, each 2 metres by 1·2 in height (about 6 feet 6 inches by 4 feet), thus dividing the box into 100 separate compartments.

The electrodes are made out of a kind of carbonaceous conglomerate, formed by ramming together a moist paste prepared by crushing and kneading together 1 part by weight of any cereal (say barley), 1 part of bituminous coal, and 2 parts of very porous clayey soil. Having been moulded and dried, these electrodes are baked in a muffle at a temperature of about 400° C. After which operation they become porous to a very high degree.

Next, one side of each piece is covered with tar, and the plate again baked, which makes that side dense and non-porous (see Fig. 104). The porous side is covered with some material like coarse canvas, and constitutes the positive side, and is depolarized by the gases that enter the space between the two parts, $a\,b$, by the holes B, the inert nitrogen escaping through the pores of a, and agitating the liquid. The oxygen and the carbon dioxide are absorbed by the liquid. The iron is put between the electrodes a, in the granulated state, or in briquettes, and the salt water is supplied by the pipes c. The action is :—Chloride of iron, caustic soda, and hydrogen are produced by the action of the salt and the iron. The hydrogen is oxidized by the air blown through, forming water. The chloride of iron is converted into carbonate by the carbon dioxide in the air, and when the water is charged with it, it is run off by the holes $d\,d$. The E.M..F. ·9 volt; current, 1000 ampères (?).

The current is taken off two metal plates, E, E', and the space between E and the next carbon plate is rammed full of coke or iron filings to conduct the heavy current.

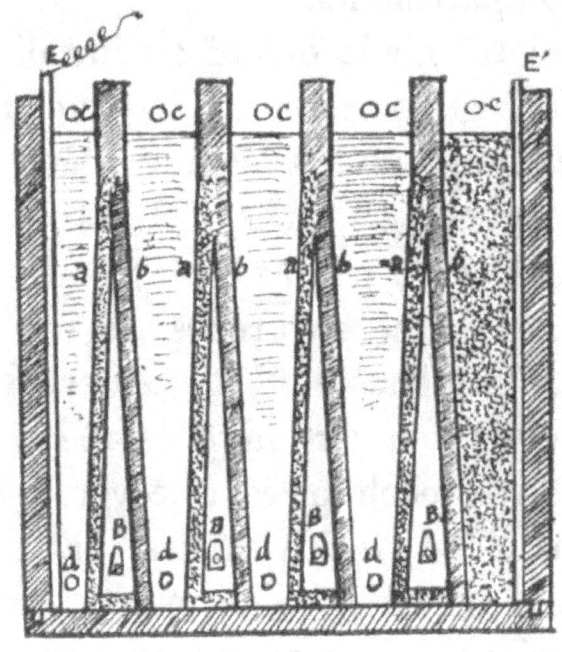

Fig. 104.

The battery is regenerative because the carbonate of soda and chloride of iron produced reaction on each other, with the result that iron carbonate is formed and precipitated, along with chloride of sodium, which is again used in the battery. The carbonate of iron may be reduced to the state of metallic iron by heating in a furnace wherein it is exposed to the action of carbon monoxide.

§ 221. *Mr. J. A. Fleming*, with a view to strengthening

the arguments in favour of the chemical theory of the action of the battery, as compared to the contact theory, devised in 1874 a battery in which there was no contact between dissimilar metals. It consisted, as shown in section at Fig. 105, in a number of test-tubes mounted in a wooden rack. Tubes 1, 3, 5, 7, 9, etc. were filled with

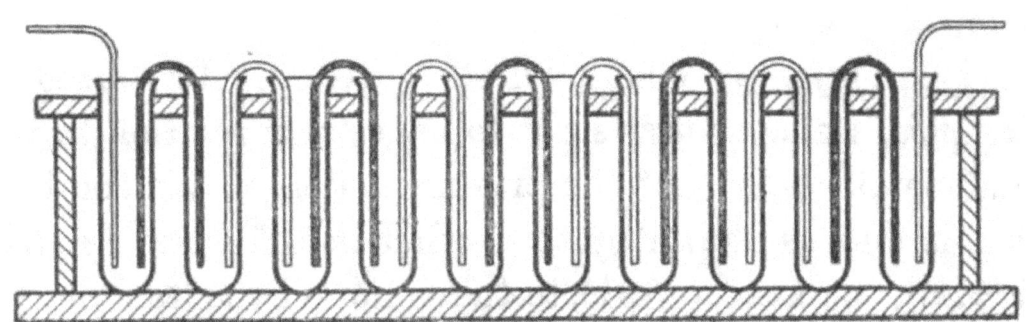

Fig. 105.

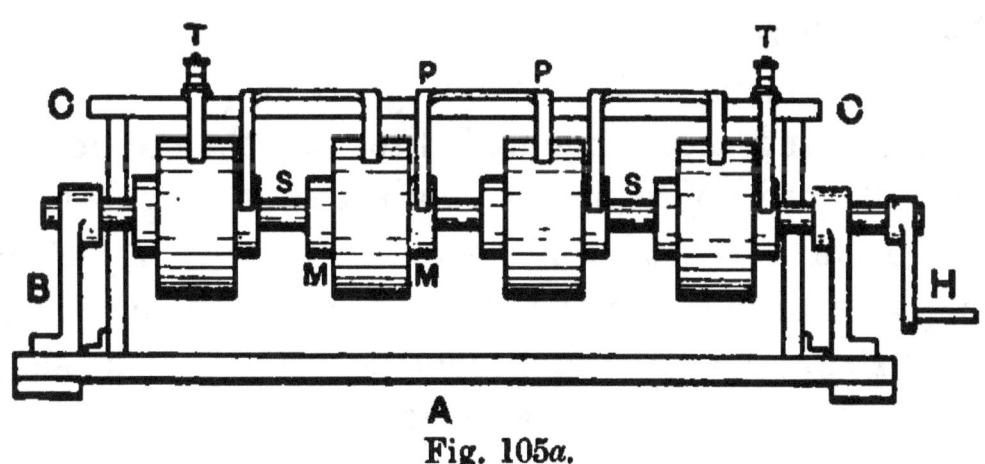

Fig. 105a.

an aqueous solution of sodium pentasulphide; tubes 2, 4, 6, 8, 10, etc. contained dilute nitric acid. Into these tubes ⋂-shaped strips, alternately of lead and of copper, dipped. In this arrangement, the same metal dips into two different menstrua, to one of which it is positive, while to the other it is negative. The two terminals are of the *same* metal. The E.M.F. of this combination is such that 60 cells were equal to 14 Daniells; hence each cell = 0·251 volt. A similar arrangement, using nitric acid throughout with the ⋂ pieces made of iron, one limb of each having been rendered " passive " by previously being dipped in strong nitric acid, acts quite as well if not better, and yet more forcibly emphasizes the truth of the chemical theory.

§ 222. *Edison's* " Chalk cylinder battery," illustrated at Fig. 105*a*, seems at first sight to have some relation to the " one metal " cells; but the primary action of this combination is due, as shown by the author in 1879 (see *English Mechanic*, vol. xxix., p. 601), to the action of the fluid on the dissimilar metal of the springs and the boss. It consists in an insulating spindle mounted on trunnions, and capable of being rotated. Four rather wide discs of chalk are mounted on metal hubs, extending about $\frac{1}{4}''$ beyond the chalk discs, and these mounted discs are then slipped on the spindle, and keyed thereto with a short gap between each. From a bar standing over the cylinder project five light platinum springs; the two outer ones being simple straight strips, the three inner ones being cut to the shape of a V. The first spring rests on the face of the chalk disc under it, one limb of the three succeeding springs lies on the metal hubs or bosses, while the other limb rests on the chalk. The last spring rests on the boss. If the chalk cylinders be moistened with a little salt water, a trifling current is set up; if the cylinders be rotated, the resistance falls greatly, and a considerable current results. The E.M.F. is about 0·5 volt for the four discs, or 0·125 per

" cell." Interesting from the lowering of resistance during rotation, which falls from 1200 ohms when still to 50 ohms when rotating.

§ 223. Another extraordinary cell, as containing no non-metallic body, is that described by *Messrs. Ayrton and Perry* in 1878. It consists of a strip of platinum and a strip of magnesium immersed in dry mercury. On first immersion the current is unsteady, but soon attains fair constancy. If the elements be short-circuited for a few minutes, and again insulated, the current becomes stronger. E.M.F. with pure dry mercury 1·56. The platinum is positive to the magnesium. It is just possible that air oxidation plays some part in this result.

§ 224. *Messrs. C. Thompson and C. R. Alder Wright,* in 1887, produced several varieties of two-fluid cells, in which one metal only was employed. Carbon may be substituted for the metal, which is generally platinum. Two different solutions are required, one on each side of the porous partition, one of which should be a reducing agent, and the other an oxidizing body. For instance, using platinum in both partitions, we may employ sulphurous acid in one compartment and a solution of chromic acid in the other; or sodium sulphite may be used against potassium permanganate. Ferrocyanide of potassium solution has also been tried against solution of chromic acid with good results. A very good cell contained a solution of lead in caustic soda on the one side, an akaline hypochlorite on the other. In another experiment, in which two carbons were used, one plunged in a mixed solution of potassium sulphide and sodium thiosulphate (hyposulphite of soda), and the other into a mixture of potassium permanganate and sulphuric acid, an E.M.F. of 1·84 volts was attained.

§ 225. Very similar to the above is *Herr Wohler's* "one metal cell." An outer vessel 6" high, 4" diameter, is fitted with a porous cell 6½" high, 2" diameter. The porous cell contains concentrated nitric acid, in which is immersed a rod or cylinder of aluminium. A larger cylinder, also of aluminium, surrounds the porous pot, standing in the outer vessel, which is filled either with very dilute hydrochloric acid, or better with a fairly strong solution of caustic soda. The E.M.F. is not high, about 1 volt; but the internal resistance is very low, so that a large current is available.

§ 226. *Riatti's* "thermal cell" also employs one metal. A wooden box, with the sides rendered waterproof by a suitable cement, is traversed by two copper tubes, one near the bottom, and the other near the top. The box is filled with a solution of sulphate of copper, kept at saturation point by a trough or bag containing crystals of sulphate of copper suspended in the solution. The two tubes are fitted with copper wires, which serve as electrodes. A stream of cold water plays through the lower tube, while a jet of steam flows through the upper one. A steady and perfectly constant current is set up, the upper tube being gradually dissolved, while the lower has copper deposited on it. Invented by Vincenzo Riatti, Professor at the Polytechnic School at Forli, Italy, 1884.

This cell leads us to the consideration of others in which *heat* plays an important part in the production or the heightening of the electrical effects.

§ 227. *Gaudini*, in 1882, constructed a cell in which an outer vessel of cast-iron contained a cylinder of zinc standing in a saturated solution of common salt. In the middle of the zinc cylinder was placed a porous cell, which contained a prism of carbon, in a saturated solution of potassium bichromate and sulphuric acid. This cell gives an intense but variable current: on heating over a gas-jet, the E.M.F. becomes higher, and the current much steadier. Cold, the E.M.F. is 2 volts, the internal resistance 0·82 ohm. When heated the E.M.F. rises to 2·44 volts, while the resistance falls to 0·71 ohm.

§ 228. *Becquerel* in 1855, *Jablochkoff* in 1877, and *Brard* in 1882, worked at the production of an electric current by the oxidation of carbon with the aid of potassium or sodium nitrate at a high temperature. The Becquerel cell consisted in a platinum crucible supported on an insulated stand. In the crucible, which had connected to it a platinum wire reaching to the outside to form one terminal, was placed a sufficiency of nitre, which was maintained in the molten condition by means of a Bunsen burner flame playing on the outside of the crucible. Suspended by a copper wire, forming the other terminal, was a prism of carbon, dipping into the molten nitre. The carbon was oxidized, with the simultaneous evolution of electricity.

§ 229. *Jablochkoff's* heat cell contains a negative electrode consisting of a block of carbon, of coke, graphite, or any artificial conglomerate of carbonaceous matter possessing the same qualities, dipping into a bath of molten nitrate (sodium, potassium, or ammonium) contained in any metallic vessel on which the melted nitrate has no action. Sodium nitrate (cubic nitre: Chili saltpetre) is preferable on the ground of cheapness. The carbon is rapidly oxidized to carbon dioxide, with the extrication of electricity. The crucible is closed by a hinged cover, having an outlet for the gases produced, and an insulating (porcelain) central perforated washer through which the carbon passes. The gases may be collected and used for the production of motive power. By mixing various metallic salts with the nitrate, not only may the E.M.F. and current be regulated, but also metallic deposits may be precipitated on the negative element. The cell can be fed with fresh carbon and fused nitrates, like a furnace, and works steadily so long as these supplies are kept up.

§ 230. *Dr. Brard* (1882) patented his " Electrogenic torch," in which a rod formed of a suitable compound of graphite in fine powder, mixed with an equivalent of saltpetre, and then kneaded and pressed into the shape of a torch, forms at once the cell and the electrodes. On applying a light to one end of the torch, the carbon enters into combination with the oxygen of the saltpetre, as shown in the annexed equation :—

$$2\,C_2 + 2\,KNO_3 = 3\,CO + N_2 + K_2CO_3.$$

Carbon and　　　saltpetre　give　carbonic and　nitrogen and potassium carbonate.
oxide

§ 231. *T. A. Edison* (1891) devised a carbon cell, which we illustrate at Fig. 106. A cast-iron vessel is fitted into a furnace. This vessel is furnished with a closely-fitting lid or cover, in which are two apertures; one central, through which a carbon or carbonaceous rod is fed, and the other lateral, to which is attached a tube connected to an aspirating fan, or other means of producing a partial vacuum. The carbon rod must be insulated by a porcelain ring or washer from the iron vessel. In this latter is placed a charge of oxide of lead, iron or copper, which is maintained at the fusing point by the heat of the furnace. The carbon seizes on the oxygen of the metallic oxide, reducing it to the metallic state with the production of carbon dioxide and the evolution of electricity. The removal of the carbon dioxide by means of the vacuum

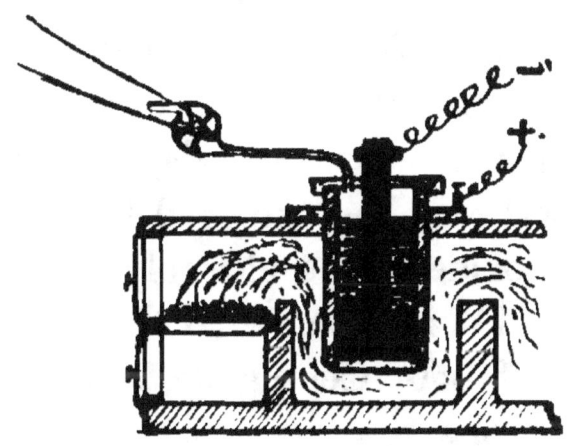

Fig. 106.

pump enhances the effects. Perhaps the best carbon-consuming cell yet produced, is that devised by Dr. W. W. Jacques, 1896.

§ 232. *Jacques'* cell consists in an outer iron vessel *I*, which should be as free from carbon as possible, furnished with a rose inlet *R* at the bottom. Attached to this inlet is a tuyère, or any other form of blast-tube, by means of which a strong current of air can be blown into the iron pot. A rod or cylinder of good conducting carbon, such as gas carbon *C*, is supported in the centre of the iron pot by the insulating cover. The iron pot is about two-thirds filled with commercial caustic soda $E\,E$; the whole is heated by a furnace to a temperature between 400° and 500° Centigrade (752° to 932° Fahr.), and the force-pump A started. This blows a stream of air through the molten soda, which, on reaching the central red-hot carbon, oxidizes it, with the production of carbon dioxide, some of which combines with the soda to form sodium carbonate, electricity being at the same time set free. Much of the carbon dioxide escapes through a vent-hole in the cover. By reason of the absorption of the carbon dioxide by the soda, and also because of the accumulation of the siliceous ash formed during the combustion of the carbon, it is necessary from time to time to draw off the spent soda, and renew it with fresh. This renewal may be postponed by the addition of a little magnesium oxide to the electrolyte. A battery of 100 such iron pots $1\frac{1}{2}''$ in diameter, $12''$ deep, maintained at a temperature of 450° Centigrade, gave a current of 16 ampères at 90 volts pressure for 19 hours right off; 30 lamps of 16 c.p. 90ᵛ being fed by them. Eight lbs. of carbon were consumed in the battery in this time; the inventor therefore claims that he gets from 80 to 90 per cent. of the actual energy contained in the carbon, as against 10 or 12 obtainable from carbon (coal) in driving a dynamo; but no account appears to have been taken of the fuel required to bring and maintain the cells up to the initial temperature of 450° Centigrade. Still, this is a very interesting cell, and one which may eventually prove of real use, since the oxidation of the carbon may be made to do useful work in the reduction of metals from their ores, etc.   E.M.F. 0·9 volt; I.R. about 0·5 ohm.   Fig. 107 is a sectional view of the Jacques' cell.

§ 233. In *Case's* "heat" cell we have another example

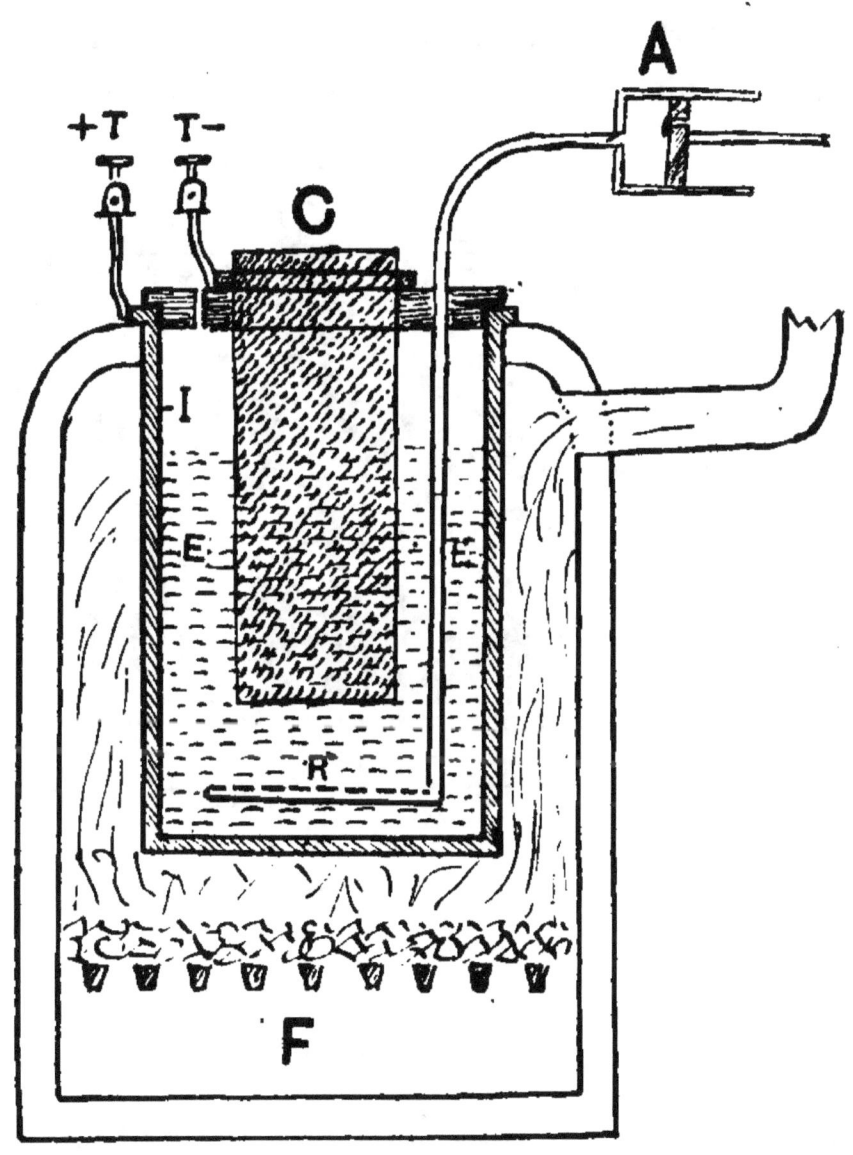

Fig. 107.

of the influence of heat in setting up a chemical reaction which liberates electrical energy.   The constituents are, carbon, tin, and a solution of chromium chloride.   This was patented by Willard E. Case, of New York, in 1886.

The arrangement is shown in our Fig. 108, where *A* is the containing vessel, closed hermetically by the cover *B*.   A carbon plate *C* rests at the bottom of this, and is connected to the terminal *T* by means of an insulated wire *D*, passing through an ebonite tube *E*.   On this lower carbon plate is placed a thick layer of finely granulated tin.

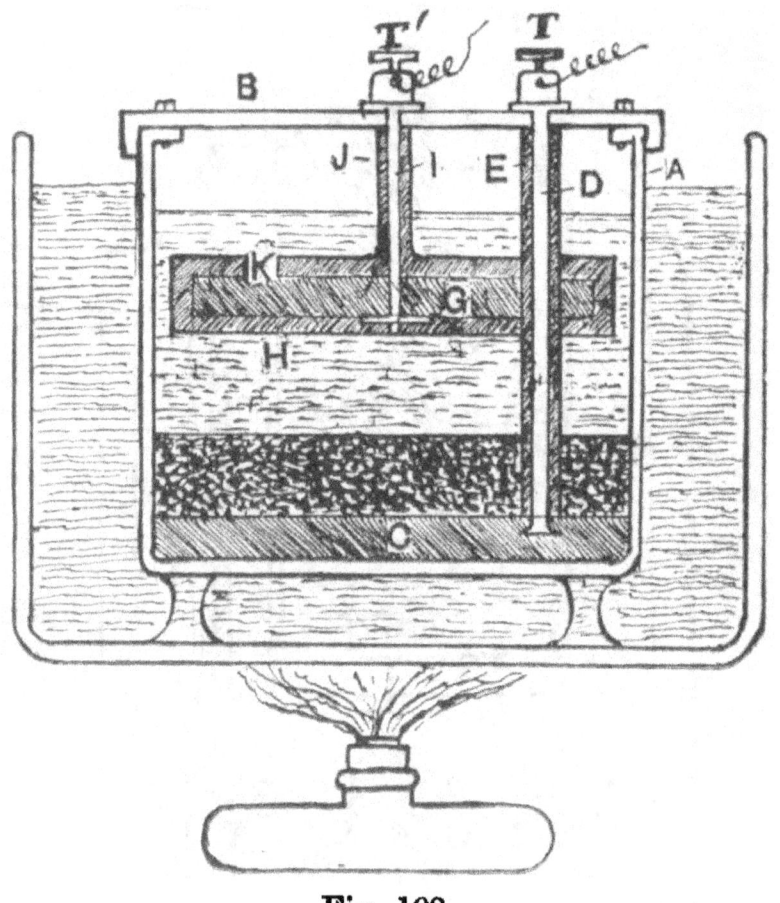

Fig. 108.

Supported above this, but not in contact with it, is a second carbon plate *G*, entirely enclosed in a covering of porous earthenware. This second carbon plate is connected to the terminal *T'* by the insulated wire *I* and *J*. The remaining space in this vessel *A* is nearly filled with a solution of chromium chloride, produced by heating together hydrochloric acid, chromic acid, and alcohol. As long as the solutions are cold, practically no current is set up, but when gradually heated the following readings were obtained on a galvanometer :—At 120° Fahr. 2°, at 130° Fahr. 4°, at 140° Fahr. 8°, at 150° Fahr. 13°, at 160° Fahr. 17°, at 170° Fahr. 22°, at 180° Fahr. 28°, at 190° Fahr. 34°, at 200° Fahr. 44°, at 207° Fahr. 49°. So as not to exceed a temperature of 212° Fahr., the apparatus is heated by being plunged in a water-bath, under which a spirit or other lamp plays, as shown in the figure. The best result, as far as regards E.M.F. and current, is obtained at about 200° Fahr. The cell, when exhausted, entirely recovers itself if allowed to cool down to about 56° Fahr. The action that goes on when heat is applied is expressed by the following equations :—(1) $Cr_2 Cl_6 = 2 Cr Cl_2 + Cl_2$. The liberated chlorine then attacks the tin, thus :— (2) $Sn + Cl_2 = Sn Cl_2$. On cooling, the action is reversed, the tin being deposited on the lower carbon plate, while the chromic chloride is re-formed thus :—$Sn Cl_2 = Sn + Cl_2$. This chlorine re-combines with the chromous chloride to form chromic chloride, thus :—$Cl_2 + 2 Cr_2 Cl_2 = Cr_2 Cl_6$.

It will be seen that the lower carbon is not active, serving only as an electrode to connect the tin to the outer terminal T. The object of encasing the upper carbon with porous earthenware is to prevent the deposition of *tin* upon its surface. This cell is perfectly regenerative.

§ 234. To *Mr. Case* we are indebted also for a carbon-consuming cell, that does not require heat to put it in action.   It is built up, as shown at Fig. 109, of an outer containing vessel *A*, a central porous pot *B*, a rod of carbon *C* (which should be selected, specially hard graphite), and a sheet, rod, or cylinder of platinum *Pt*.   The carbon rod in the porous cell is packed tightly round with granular carbon, previously soaked in oil of vitriol.   The containing vessel *A* is ¾ filled with sulphuric acid, strength about 1 part of pure acid to 3 parts of water.   In this condition no galvanic action takes place.   If now, powdered chlorate

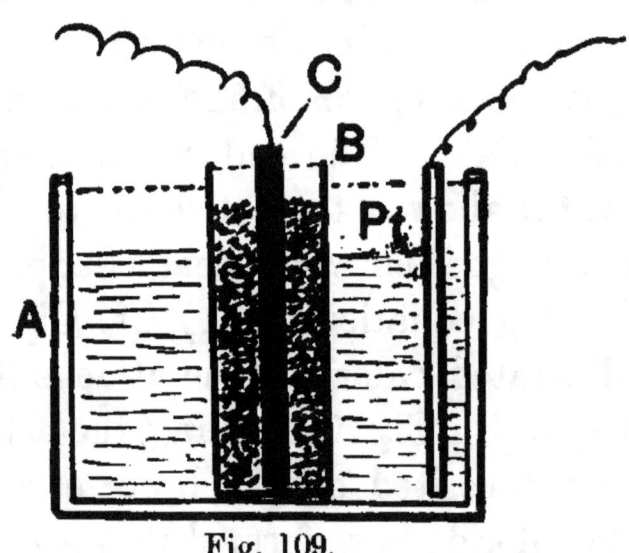

Fig. 109.

of potash be gradually added to the sulphuric acid in the outer vessel (and this had better be done in a darkened room, as sunlight *may* cause an explosion of the peroxide of chlorine evolved), the liquid becomes *red*, owing to the separation of this body.   The peroxide of chlorine thus liberated, permeates the porous cell, and reaching the crushed carbon therein contained, is deoxidized by it, yielding up its oxygen, and giving rise to carbon dioxide and chlorine gas.   In other words, the carbon in the porous cell is literally burnt in the oxygen contained in the chloric peroxide Cl O_2, thus $C + Cl\, O_2 = CO + Cl$. As the fluid in the outer vessel becomes decolorized, more chlorate of potash should be added. E.M.F. 1·25 volt.

§ 235. The couple described by *C. J. Reed*, in 1896, under the name of "thermo-tropic cell," has some resemblance to the well-known thermopile; but it differs therefrom in the fact that the current set up is due to the decomposition of an electrolyte, under the stimulus of heat. For its original form, it consists of a U of stout copper wire. This is clamped at the free ends of the limbs, between two flat pieces of wood, leaving the bent portion free. This portion is now severed, and the freshly-cut portion held over the oxidizing portion of a gas or ether flame until coated with a film of oxide; or a solution of any akali may be rubbed over the cut surfaces, and dried thereon. These cut extremities are then sprung together so as to make contact. A wire is now attached to each of the straight ends of the U, to serve as electrodes. On heating *one* of the wires near the junction, a strong current is set up, provided the other wire be kept cool. The E.M.F. is from 0·2 to 0·4 volt. The oxide of copper is reduced to metallic copper on one wire, during this action.

§ 236. There is an intimate relation between the power possessed by heat of favouring certain chemical reactions, and thus setting up electrical currents, and that of *light*. *Professor Grove* noticed the effect of light falling on one of the platinum electrodes of a voltmeter, and mentions the appearance of a current therefrom in his *Correlation of Forces*. Becquerel and Robert Sabine also noticed these effects; but the first one to produce a real " photo-electric cell " was *Professor Minchin*, about 1880. Two sheets of tinfoil separated by a sheet of blotting-paper, each sheet being furnished with a terminal, are immersed in a test-tube containing plain water. If while one sheet is kept shaded, the other be exposed to a bright light, as sunlight or that of burning magnesium, a current is set up, as shown by a galvanometer placed in circuit.

Professor Fleming modified this cell, by forming the two tinfoil sheets into two cylinders, the inner one being the smaller, enveloped in a roll of blotting-paper, and surrounded by the larger tinfoil cylinder, water as before being the fluid. This construction does away with the necessity of shading one sheet of tinfoil, since the outer cylinder and the blotting-paper effectually shade the inner one.

Kitching describes a cell constructed by him (*circa* 1882), in which two plates or strips of silver, coated with freshly precipitated silver chloride, and immersed in any transparent vessel containing water slightly acidulated with hydrochloric acid, gives a current when one of the plates is exposed to light, the other being darkened. In this case a current is set up, which flows *from* the light plate to the one which is shaded.

§ 237. *Mr. Saur's* photo-electric cell, or "impulsion cell" (1882), appears to be the best of this class. The constituents are sulphide of silver, in common salt and sulphate of copper on the one hand, and platinum in mercury, on the other. A square glass vessel standing in an opaque box, contains a porous pot, in which is placed a platinum electrode, dipping into mercury lying at the bottom of the pot. The outer vessel contains a mixed solution of 15 parts of salt, 7 parts of copper sulphate, and 100 parts of water. Into this dips an electrode made of silver sulphide. On first closing the circuit, through a galvanometer, while the opaque box is entirely closed, so that all light is excluded, a deflection shows the presence of a current, the platinum being the positive element. If now light be admitted, by opening one of the sides of the opaque case, the current is largely increased, and that in proportion to the intensity of the light. This cell is so sensitive to any variation in light, that even a passing cloud causes a variation in the current set up. Hence this cell in connection with a galvanometer, a tracing-point and an unrolling band of lined paper, has been suggested as a light-recorder. The action of the cell is explained as follows :—

1st. The formation of cupric chloride by the action of the salt upon the copper sulphate, thus :—$2 \, Na \, Cl + Cu \, SO_4 = Na_2 \, SO_4 + Cu \, Cl_2$.

2nd. The mercury is attacked by the copper chloride, with the production of mercuric chloride and cuprous chloride, $Hg + 2 \, Cu \, Cl_2 = Cu_2 \, Cl_2 + Hg \, Cl_2$.

3rd. The reduction of the silver sulphide under the combined influence of light and cuprous chloride :—

$$Ag_2 \, S + Cu_2 \, Cl_2 = 2Ag \, Cl + Cu_2 \, S.$$

This last reaction requires solar light to effect it, hence the peculiar behaviour of the cell. Fig. 110 A shows the cells with the case closed ; B represents one side of the case open, so that light can fall on the silver sulphide plate *Ag*.

Just recently, Riggollot and John A. Randall have experimented and perfected another form of "light" cell, in which copper oxide plates, immersed in plain water, form the active elements. On exposing one of these plates to light, an electric current is set up, which can be detected by the galvanometer. The size used by Riggollot for his plates was $\frac{3}{4}''$ long by $\frac{1}{4}''$ wide. The action of these cells is very regular, and increases (up to a certain point) with the intensity of the light. Thus, if one candle at 1 foot distance gives an E.M.F. of 1 unit, 6 candles will cause the cell to give 6 units, and so on. Using magnesium wire, gas, or diffused daylight as the illuminant, the increase in E.M.F. is also uniform, though it is questionable, whether with more powerful sources of illumination the E.M.F. is proportional to the intensity. These cells may be used continuously without any falling-off in sensitiveness. To increase the E.M.F. it is necessary to couple up several cells in series. As far as J. A. Randall's experiments go, they would tend to prove that the E.M.F. is proportional to the number of cells. Each cell in ordinary conditions gives about $\frac{1}{1000}$ of a volt. One peculiarity, noticed also by Minchin, is that the E.M.F. is increased largely by dyeing the plates with certain dyes. Thus, Riggollot got the best effects when the plates were dyed with "malachite green" (an aniline dye), while Randall was more successful with eosin and naphthol yellow. The E.M.F. rises from $\frac{1}{1000}$ of a volt to $\frac{1}{20}$. Another peculiarity is, that if the plate exposed to light be shifted along the spectrum, it will be + at the red end, but change to − before reaching the violet.

§ 238. The *English Mechanic* coal-gas cell (1879). This is constructed with a 3 lb. jam-jar, in which is inserted a sheet of lead 6″ × 8″ bent round so as to line the inside. A lug from this lead is brought over the edge of the jar, to which is affixed one terminal. Inside this lead-lined

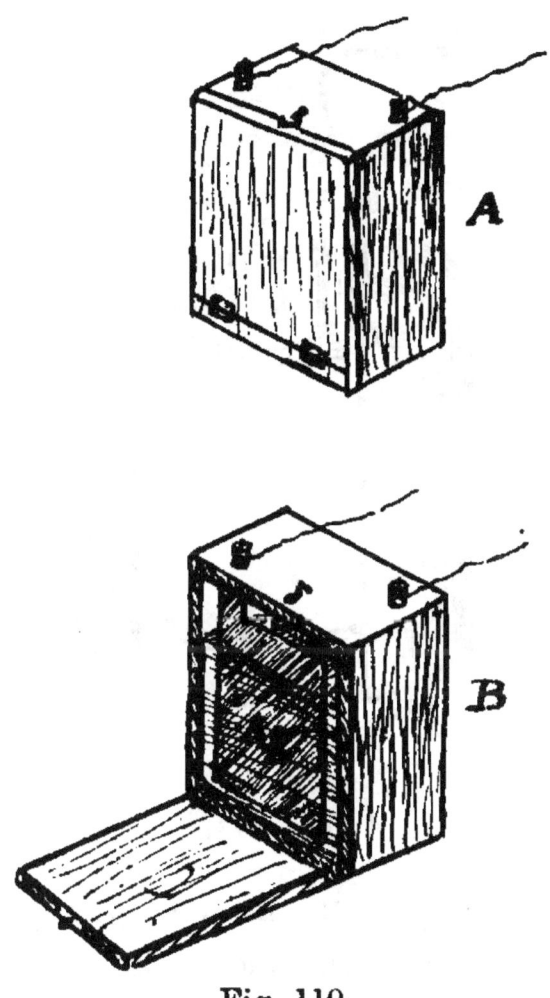

Fig. 110.

jar is placed an 8″ gas-burner chimney, fitted with a mahogany top, which must be made to fit accurately. Inside the glass chimney is also a roll of lead, made from a piece of sheet-lead 8″ × 6″. A brass or copper rod is soldered to the upper extremity of this second cylinder of lead, and passing through the mahogany cover, forms the other terminal.   The outer jar has about 2″ deep of dilute sulphuric acid placed in it, the inner glass chimney is filled with coal-gas by holding over a gas-jet, it is then replaced in the outer jar, and all interstices filled in with warm paraffin wax.   We therefore have practically lead in

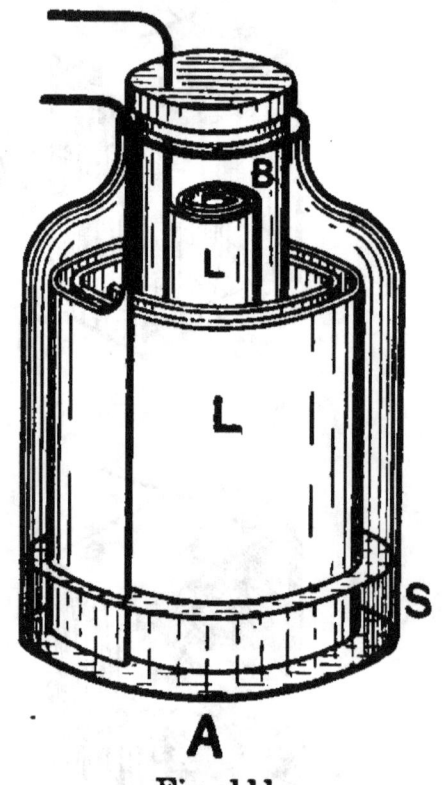

Fig. 111.

air on one side, and lead in impure *hydrogen* on the other. A current is set up with the production of *water* and *carbon dioxide*.   Fig. 111 shows the general construction.

§ 239. *J. A. Kendall's* thermal gas cell.   Constituents are two platinum tubes, a stream of oxygen and of hydrogen gases, and a source of heat.   A platinum tube, sealed at the lower extremity, is fitted with a smaller platinum likewise sealed at the bottom.   The space between

the two tubes is filled with glass, which is kept in a
molten condition by means of a furnace, or Fletcher's
burner.   Into the inner tube a stream of hydrogen is continually pumped, while oxygen or air is supplied to the
outside.   Platinum wires attached to the inner and outer

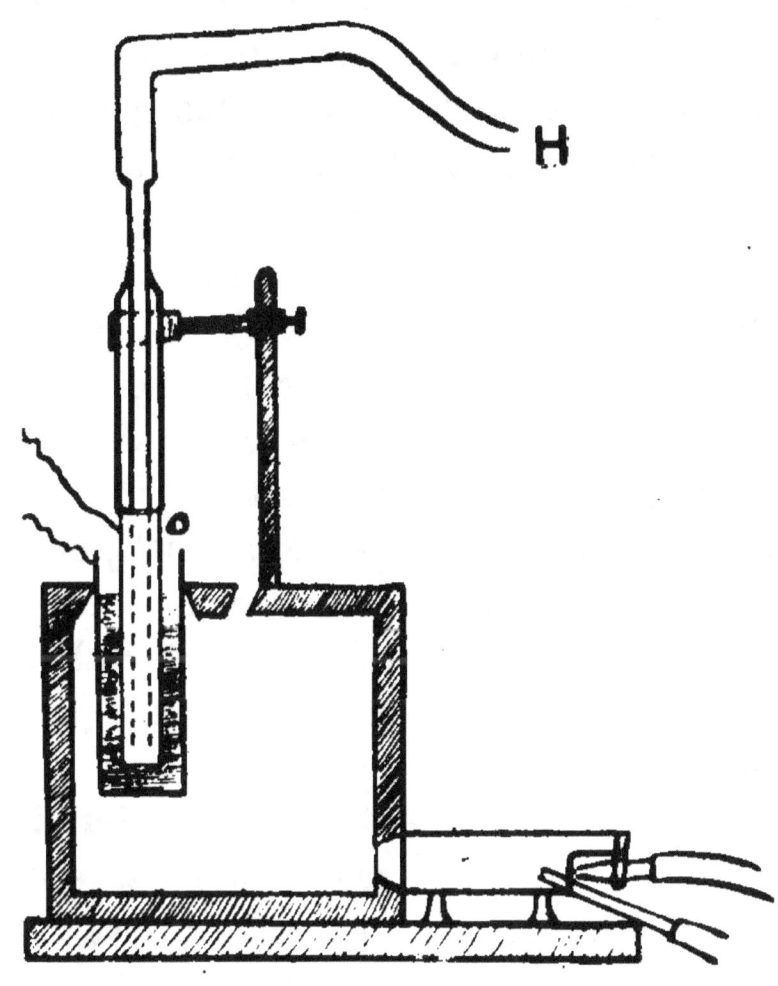

Fig. 112.

platinum cylinders form the electrodes.   On heating the
outer cylinder to bright redness by means of reducing
flame, so that the glass is melted, the hydrogen rapidly
permeates the pores of the inner platinum tube, and,
passing through, enters into combination with the oxygen,

giving rise to an E.M.F. of 0·7 volt. In the apparatus shown at the Inventions Exhibition in 1885, which gained the inventor a silver medal, the outer platinum cell was $2\frac{3}{4}''$ deep and $1''$ diameter, the inner being rather longer, and only $\frac{5}{8}''$ diameter, the portion immersed in the fused glass being $2\frac{1}{4}''$ long. In these circumstances it was found that the hydrogen gas passed through the inner platinum tube, at the rate of about 5·32 cubic centimetres per hour. If the tubes be heated by an oxidizing flame no current is set up, and no hydrogen flows through the platinum. Fig. 112 is a sectional view of a single cell.

§ 240. *Scharf's* gas cell (1890). Constituents :—Porous carbon plates, acidulated water, hydrogen and oxygen (see Fig. 113). Arrangement :—A metal holder A, preferably cylindrical, provided with a bottom and cover, $A'\ A''$, by means of which it can be hermetically sealed, by screws or otherwise. In this are the cells, formed each of two electrodes, the latter forming an interval. The inventor used porous carbons. Each electrode is enclosed by a metal ring having a projection at one part. This ring is to connect the electrode with the terminal wires. In order to keep the electrodes at suitable distances apart there is placed between each pair an insulation e of ebonite or pressed paper, or better asbestos, fitting accurately and closely into the outer holder A. Every alternate interval or space between every two cells is filled with a liquid conductor, such as acidulated water. These intervals may be fitted with porous non-conductors saturated with the electrolyte.

All the even cells contain one gas, and all the odd

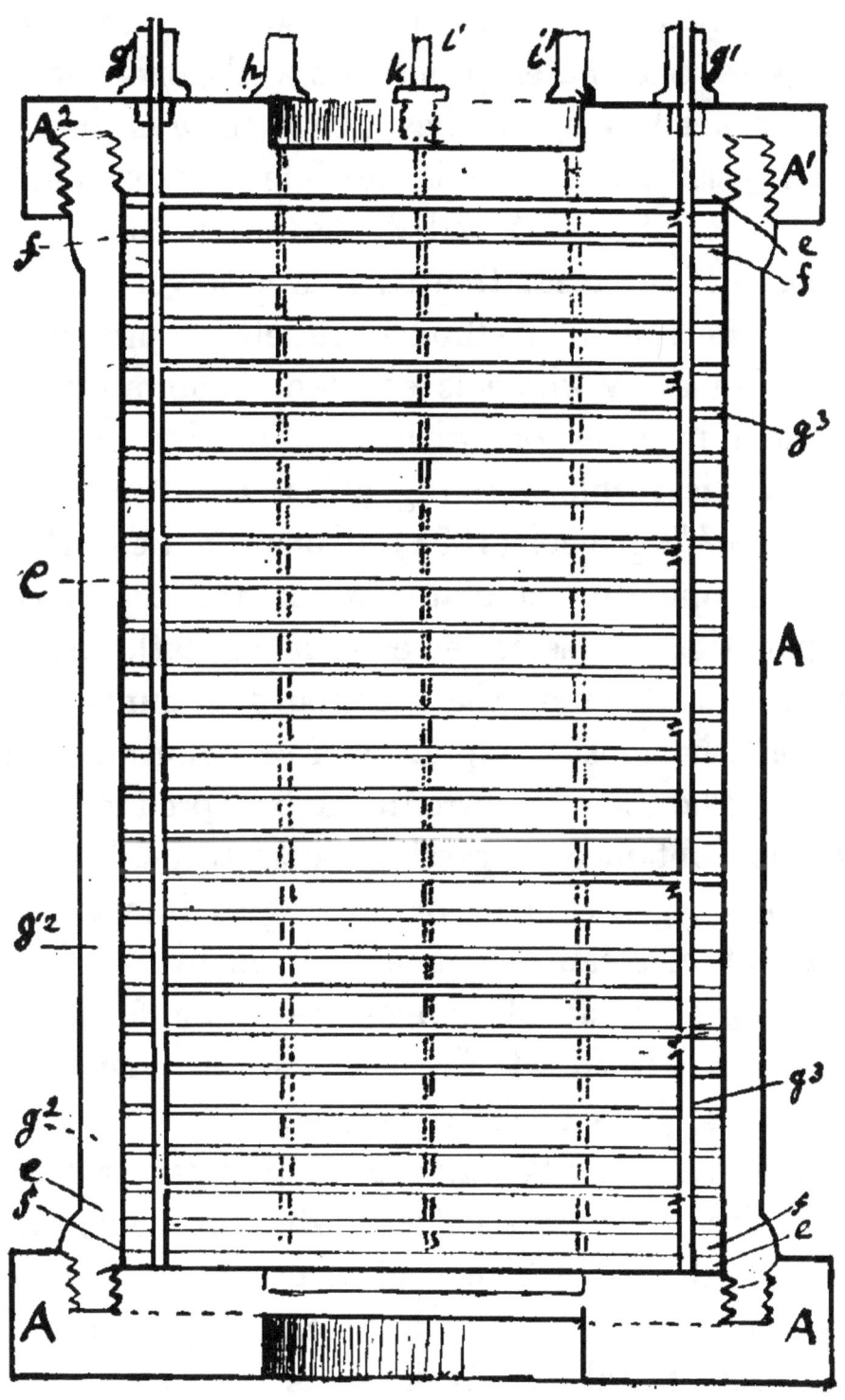

Fig. 113.

ones contain the other. All the like cells are connected together electrically. For the introduction and removal of the gases and water from the cells, one end of the outer container has eight holes. Of these g is the inlet for hydrogen, g' is its outlet; h is the inlet for oxygen, and h' its outlet; i is the inlet and i' the outlet for acidulated water. Each of the insulating rings E is traversed inwardly by one of the influx or efflux channels, or provided with a notch E' or e', by which is effected a communication of the channel in question with the interior of the cell formed by the insulating ring affected.

§ 241. *Dahl's* gas cell (1891). Constituents:—Platinum electrodes dipping in acidulated water, and separated by a porous partition whose lower edge is beneath the surface of the liquid. One of the electrodes is surrounded by oxygen, or other gas or liquid such as chlorine, bromine, or iodine. The other is surrounded by "producer gas" or "water gas," obtained by passing steam through white-hot coke. The oxygen is made by heating sodium manganate in a current of steam, when oxygen is liberated, caustic soda and lower oxide of manganese being formed. On heating in a stream of air, the sodium manganate is reproduced, so it is never consumed.

§ 242. *Mr. Andrew Plecher*, a German inventor, has lately (1901) devised a new gas battery, which presents some interesting peculiarities. The electromotive force is produced by the combination of hydrogen and oxygen, this being brought about by taking advantage of the well-known properties of finely-divided platinum. Figs. 114 A and B show the section and exterior view of this battery, which consists of a series of chambers C, made of a specially prepared material, for which purpose a mixture of clay or plaster with a solution of chloride of platinum is generally used.

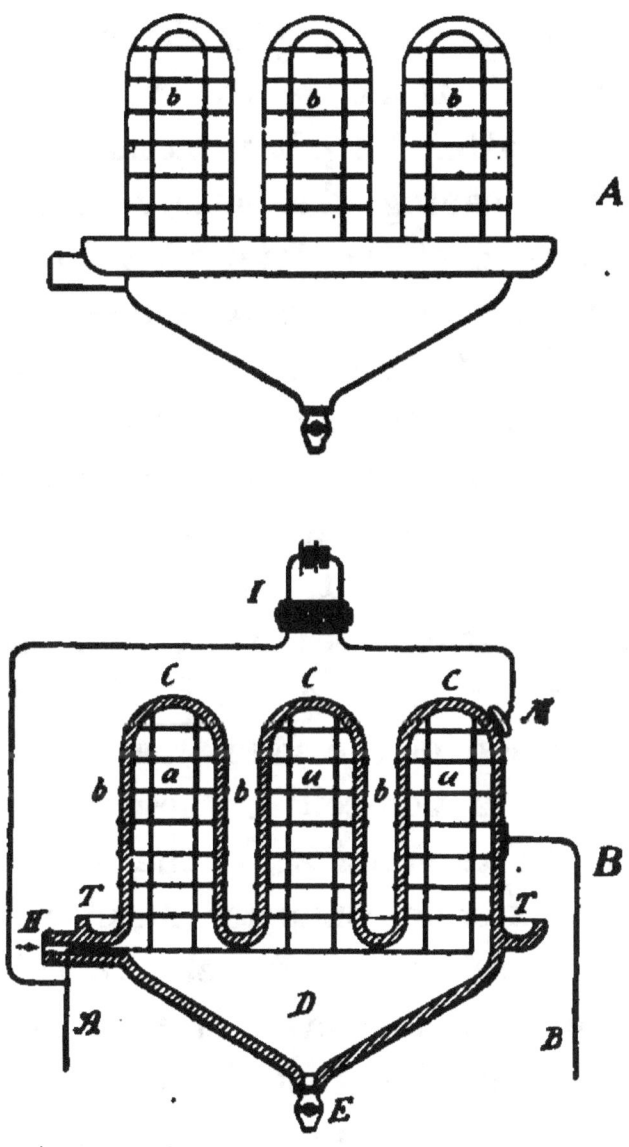

Fig. 114.

The vessels are dried and then baked, so as to harden the material and decompose the platinum salt, when the metal assumes a finely divided state. In the interior of the vessel is a system of conductors a, formed by rods or wires

x

attached to the walls; the conductors of all the vessels are connected to a common terminal *A*. On the exterior of the vessels is placed a similar system *b*, connected to the terminal *B*. The vessels rest upon a lower portion *D*, with inclined walls, and having at the bottom an escape-cock *E*. At the side is an opening *H*, for the admission of the gas. The battery is put in action by introducing hydrogen by this opening, the battery being filled with gas, while the oxygen at the exterior is furnished by the surrounding air. The two gases penetrate into the porous walls of the vessel, and combine under the influence of the finely-divided metal; water is formed by this combination, and this runs down the sides of the interior, and is taken out at the bottom, while that formed at the exterior of the vessels is collected in a trough *T*, surrounding these. This combination of the gases sets up an electro-motive force between the terminals *A* and *B*. It has been found that the combination of the gases may be increased by the action of an induction-coil *I*, whose secondary circuit is connected to the terminal *A*, on the one hand, and on the other to an exterior conductor *M*, placed at a short distance from the walls of the vessel. In this battery no electrolyte is needed; it is even best to get rid of the water as much as possible, and its evaporation may be hastened by a current of air. Palladium may be used instead of platinum, as its action is similar in this respect. The action is not confined to hydrogen and oxygen, but other gases may be used.

CHAPTER VII

DRY PILES AND DRY CELLS

§ 243. THE inconvenience of moving about cells containing liquids, more especially when the contents were of an acid or corrosive nature, was early felt. The attention of electricians was therefore soon directed to the solution of the problem of producing a cell or battery, which, although not containing fluids that could easily be spilt, should be capable of furnishing a current of sufficient ampèrage to be of use for practical purposes, with a fair amount of constancy. The first attempts were by no means satisfactory, as they gave results that were more interesting from a scientific point of view than from that of actual utility. It is true that one of the first " dry piles " found an application in the production of a sensitive form of electroscope, known as "Bohnenberger's" from the name of the deviser, and that such a pile has also been employed for the ringing of electrical chimes over lengthened periods of time; but as the energy obtainable from any cell is always in direct proportion to the consumption of its materials in the way of chemical action, it follows that the current available in these first dry piles, in which the chemical action was extremely limited, was also exceedingly small.

§ 244. The first dry pile was the invention of *M. de Luc* (about 1804), who gave it the name of the "Electrical Column." It was constructed of discs, about 1″ in diameter punched out of paper which had previously been silvered on one side, alternating with similar discs of very thin sheet zinc. It is essential that the silvered surfaces of the paper disc should be all turned in the same direction, so that the alternations may practically be silver, paper, zinc—silver, paper, zinc, and so on. A glass tube, rather over 1″ in internal diameter, and capable of being fitted with brass caps or heads, was filled up with the discs, in the above-mentioned order, and the caps (which must come into metallic contact with the zinc at the one end and the silver at the other) placed on the tube, and cemented in place with sealing-wax. A very large number of discs can be contained in a tube of moderate dimensions. The largest instrument of this kind was one constructed by Mr. Singer. It consisted of 20,000 discs. Each of the two ends or poles affected the electrometer, and exhibited electrical attractions and repulsions; it also gave minute sparks and shocks of considerable power, but owing to the high internal resistance it was impossible to effect chemical decomposition by its aid, when applied to fluids in its circuit.

If two upright electrical columns be placed side by side, with their poles in opposite directions, and connected metallically at their upper ends, while a small bell is attached to the lower end of each, the whole will act as one column, and each bell will in consequence of the electrical actions be alternately struck by a brass ball suspended between them, and thus a continual ringing will be produced as long as the apparatus remains in action, which is generally for a year or more. This action is, however, kept up solely by the presence of moisture in the paper, for it does not take place at all when the paper is perfectly dry, and although the process of oxidation is very slow, the more oxidisable metal is in process of time found to be tarnished. An apparatus somewhat analogous to that of De Luc was constructed by *Hachette* and *Desorme* with pairs of metallic plates, separated by layers of farinaceous paste mixed with common salt. To this instrument, although it evidently owed its efficacy to the moisture of the paste, they gave the very inappropriate name of " dry pile."

§ 245. *Zamboni's* dry pile. Constituents :—Discs of paper coated on one side with tin or silver, and on the other side with manganese dioxide mixed with honey. The manganese dioxide is rubbed on the moistened paper with a cork. The discs are generally about 1″ in diameter. Arrangement :—The discs are piled together in a glass tube, in the order—tin, paper, manganese dioxide. Dry piles are remarkable for the permanency of their action, which may continue for several years. Their action depends greatly on the temperature and humidity of the air. It is stronger in summer than in winter, and a strong heat revives it when it appears extinct. A Zamboni's pile of 2000 couples gives neither shock nor spark, but can charge a Leyden jar or other condenser; a certain time is, however, necessary to this result. With a pile of 2000 to 20,000 couples, electric bells may be rung continuously for months together, electric sparks produced, Leyden jars charged, chemical decomposition effected, and many other striking phenomena caused.

§ 246. *Prince Bagration* (1847) invented a dry cell which ·is also known as Professor Jacobi's, as this latter introduced it to public notice at St. Petersburg. It consisted of any vessel impermeable to water, such as a jam-pot, a flower-pot, with the bottom hole plugged, and filled with earth saturated with a strong solution of ammonium chloride in water. Imbedded in the soil are a plate of copper and a plate of zinc, as large as possible, and at a moderate distance apart. It is advisable to prepare the surface of the copper by plunging it into a solution of sal ammoniac in water and allowing it to dry, repeating this until decided oxidation shows itself on the surface. Thus prepared, it is fairly constant in action. E.M.F. about 1 volt; I.R. about 10 ohms. A modification of the Jacobi cell, in which the zincs and coppers were arranged back to back to form divisions in a box, like Cruickshank's trough battery, moistened sand being used between the partitions, has been used for telegraphic work, etc.

§ 247. *Skrivanow's* dry cell (1882) consists in a plate of carbon, a plate of zinc, carefully amalgamated, and a special exciting paste. This paste consists of ammonio-chloride of mercury ($Hg\ CI_2$, $NH_4\ Cl$) 18 parts by weight, and sufficient solution of zinc chloride to make up a stiff paste. To prepare the ammonio-chloride of mercury, boil for three hours in a porcelain dish a concentrated solution of pure ammonium chloride with a quarter of its weight of mercuric oxide, which will form a greyish-white precipitate: Wash this, dry, and coarsely pound it, which will give a salt which is durable, unimpaired by moist or dry air, and easily preserved. To this powder add a slightly acid solution of zinc chloride of about 50° Beaumé, until a consistent paste is formed. Carefully triturate in a porcelain mortar with the previously mentioned proportion of common salt, and mould into cakes of any desired form or size. The plate of carbon, previously treated with paraffin, at the upper extremity, is covered with a cake of the paste from 1 to 3 millimetres (0·04 to 0·12 inch) in thickness, after which from 5 to 10 turns of Swedish filtering paper, or of asbestos cloth, are rolled round it. Either of these must be soaked in a saturated solution of zinc and sodium chlorides, to which a little glycerine may be added. (This is hygroscopic, and retains the moisture.) Round this, and in contact with the packing, comes the zinc, which may take the form of a cylindrical pot, a plate, or a square box, at will. Or the zinc may be placed in the centre, the carbon cylinder outside, in a suitable casing, and the paste rammed in between the zinc and the carbon. In any case the cell should be fitted with a cover through which the electrodes protrude. E.M.F. 1·5 volts. This comes very close to the dry cells of to-day. The nearest one, however, was :—

§ 248. *Becquerel and Onimus'* "Experimental cell" (1884). This consists of zinc and carbon standing in a made paste with plaster of Paris, a solution of zinc chloride and sal ammoniac in water mixed with sesquioxide of iron and peroxide of manganese.

§ 249. Shortly after this date *Gassner* placed the first

commercially successful dry cell upon the market. This consisted in a cylindrical zinc vessel about 7″ in height and 3″ in diameter, the interior of which was coated to within about an inch of the edge with a layer of plaster of Paris about $\frac{1}{4}$″ thick. Sal ammoniac, zinc chloride, and a little glycerine are to be added to the water with which this plaster paste is made up. When the coating of plaster (which must cover the bottom of the zinc cylinder to the same depth, viz. $\frac{1}{4}$″) has set, a carbon rod, about 1″ diameter and 7″ long is stood upright in it, and the space between the carbon and the plaster-coating filled in by ramming with a mixture consisting of 1 part of granular manganese dioxide, 8 parts coarsely powdered graphite, $\frac{1}{2}$ part zinc sulphate, and $\frac{1}{2}$ part of ammonium chloride, along with sufficient water to make a stiff paste. (It is advisable to add a $\frac{1}{8}$ part of glycerine, as this causes the mixture to retain and absorb moisture.) When the cell has been filled to within about 1″ of the top, the edges of the zinc and the top of the carbon are carefully wiped, and the cell sealed up by pouring in melted pitch or other resinous mixture. To allow the escape of gases generated, two short pieces of fine glass tube, reaching from the black paste to the level of the zinc pot, are pushed into the black paste previous to pouring in the melted pitch. A wire is soldered to the outside of the zinc cylinder to form one electrode, and a terminal screwed into the carbon rod to form the other. To prevent accidental short-circuiting, the zinc cylinder is usually enclosed in an outer paper or cardboard covering. Fig. 115 illustrates the original form of the Gassner dry cell.

The results obtained from these cells was so good,
being equal if not superior to the Leclanché, that
many turned their attention to the manufacture of
similar cells, with such modifications in the composition of the pastes as fancy dictated or scientific knowledge indicated.   The E.M.F of the Gassner cell was
1·5 volts, and the internal resistance in the 7″ × 3″ size
about 0·5 ohm when fresh.   But the plaster paste soon
got hardened, and the internal resistance increased.   Hence

Fig. 115.

other substances were soon employed instead of, or in
conjunction with, plaster, to form the "white paste" or
zinc lining.   The *Electric Stores Co.* produced a cell in
which fine sawdust, previously soaked in mixed chloride of
zinc and chloride of ammonium solution, took the place of
the white paste.   The *Obach* and *Hellesen* cells of Messrs.
Siemens Brothers are examples of very high-class cells,
and were made known about 1890.   In these the tendency
of "blowing" the cases, due to the evolution of gas, is
almost entirely eliminated by care in constructional details.

The Hellesen cell (Fig. 116) consists in a zinc vessel standing in a second case, somewhat larger. Both these zinc cases are perforated; the inner near the top, and the outer near the bottom. The space between the two zinc cases is filled with slagwool or fine sawdust. The central carbon rod has, at its lower half, a muslin bag tied round it; this bag being filled with the depolarizing paste (black oxide of manganese and carbon, etc.). The remaining space

Fig. 116.

between the carbon and the zinc is filled up to within half-an-inch of the perforation in the inner zinc case with a slimy jelly (gelatine or agar-agar) containing the exciting salts (sal ammoniac, chloride of calcium, chloride of zinc, etc.). Over this is placed a layer of plaster of Paris, reaching somewhat above the level of the perforation in the inner zinc, and lastly comes the layer of melted pitch. The inner zinc is fitted with a soldered wire, to serve as a terminal: to the carbon is affixed a binding screw for the same purpose. Any gases evolved force their way through the plaster layer, escaping by the holes in the inner zinc, thence proceeding downwards through the slagwool packing, and finally escaping through the perforations near the bottom of the outer zinc case. The result of this mode

Fig. 117.

of construction with the gelatinous excitant is to produce a cell having an E.M.F. of 1·5 volts, with an internal resistance of only 0·2 ohm in a cell $7\frac{1}{2}'' \times 3'' \times 3''$.

§ 250. The "*Obach*" cell of the same firm is almost precisely similar in general construction, but owing to the presence of a more active excitant in the jelly next the zinc the resistance is somewhat lower (about 0·15 ohm), and the E.M.F. somewhat higher (1·56 volts). Both these cells are noted for their staying powers. Cases are on record in which an Obach cell, starting with an E.M.F. of 1·56 volts and 0·15 ohm internal resistance, has run for 30 months, doing telegraphic work daily (Sundays excepted) from 10 a.m. to 9 p.m., when the voltage had fallen to 1·31, and the resistance had risen to 1·7 ohms.

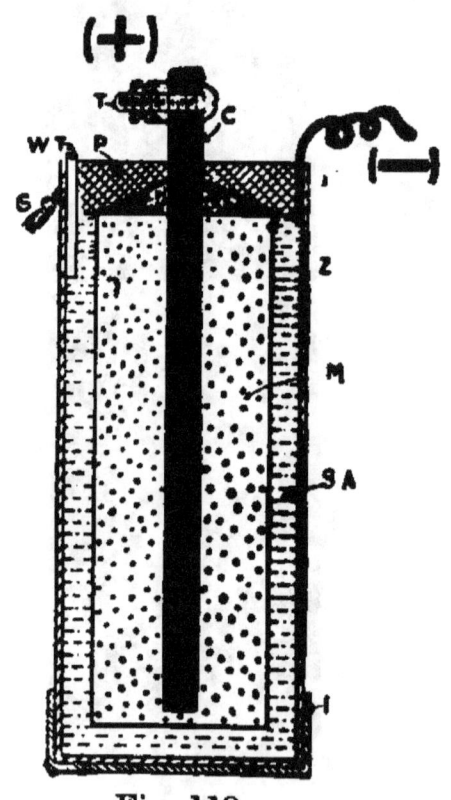

Fig. 118.

§ 251. The *Burnley* dry cell (about 1890), which was put on the market by the General Electric Co. under the name of the "E. C. C.," is another excellent cell, of which Fig. 117 illustrates the general outward appearance, and Fig. 118 gives a sectional view, illustrative of the arrangement of the interior. The composition of the pastes employed, as certified by an analysis performed by Dr. Henderson, of the Glasgow and West of Scotland Technical College, is given below :—

Analysis of Black Paste (*M* in Fig. 118):

Manganese dioxide	7·57
Carbon	47·26
Magnesia	7·20
Lime	1·40
Ferric oxide	2·12
Ammonia	12·78
Water, etc.	21·67
	100·00

Analysis of White Paste (*L* in Fig. 118):

Lime [1]	20·89
Ammonia [1]	20·29
Chlorine	6·31
Water and binding material	52·51
	100·00

In a bell-ringing test, during which contact was made for six minutes at a time, with a corresponding interval of six minutes' rest, a No. 2 E.C.C. cell kept the bell ringing for 498 hours 20 minutes before it was so far exhausted as to be unable to ring the bell. This corresponds to 249 hours 10 minutes' actual work, with a corresponding time of rest. The E.M.F. of this cell was about 1·48 volts, internal resistance about 0·3 ohm.

§ 252. The "*Carbone Co.*" makes also very good cells.

---

[1] Probably in the form of chlorides.

The general principle of construction is the same as the two latter, but their cells in the larger sizes *A* to *D* (see Fig. 119) are enclosed in outer vitrite cases, and the carbons are fitted with special unoxidizable contacts.   A 7″ × 3½″ cylindrical cell has an E.M.F. of 1·5 volts with a resistance of only 0·22 ohm, hence it can give as much as 6·8 ampères on the short circuit.   The recuperative power of the

Fig. 119.

Carbone cell is also very great.   This firm also makes another type of dry cell, in which the excitant paste is more active, and consequently gives a higher E.M.F., viz. 1·6 volts, the internal resistance being only 0·16 in a square cell, the over all size of which is 7⅜″ × 2⅞″ × 2 7/16 ″. Such a cell, which is called "The Sans Pareil," can give therefore 10 ampères on the short circuit.   It is eminently adapted to the electric ignition of motor-car engines.

§ 253. During the last few years this type of cell has been much improved, and is far superior to those that were procurable some years ago. Of recent years these cells have found much favour amongst electricians for open-circuit work in place of the Leclanché cell, because of several advantages they offer over cells of the latter type. They are much more convenient to handle, are cleaner, as they have no (free) liquid to spill and corrode instruments, etc., to which they may come in contact, they take up less room, are supplied ready charged, and have a lower internal resistance than the Leclanché, which fact also widens their field of application. Added to these advantages these cells (containing no liquid) may be used in any position, do not require water to be added to make up for the loss due to evaporation, and finally, the life of a good dry cell is quite sufficient to give satisfaction.

In principle, dry cells are similar to the Leclanché cell, and are applicable for all purposes for which that cell is employed; their behaviour in use is also similar, for, like it, their E.M.F. falls (although not nearly as rapidly) when the circuit is closed for any length of time. Also, like the Leclanché cell (the construction of which has been described in a previous chapter), their positive and negative elements are zinc and carbon, and their depolarizer and excitant are manganese dioxide and sal ammoniac solution, the main difference being broadly that the latter instead of being in the form of a liquid is combined with plaster of Paris and flour in the form of a paste, or absorbed by sawdust or equivalent material, whichever plan is favoured by the particular maker, a deliquescent salt always being added in order to prevent the drying of the paste. Among the many purposes for which dry cells are now largely used, may be mentioned telephone work (the leading telephone companies having adopted them exclusively), telegraph work (both ordinary and wireless), electric clock work (one of the most successful electric clock syndicates practically using them exclusively), for electric bells and

Fig. 120.

alarms, for intermittent use with small glow-lamps, for the electric ignition of motor-cars, the working of medical and induction coils, and the small sizes for constant-current medical batteries. In designing apparatus for use with dry cells, it is advisable, in order to insure economical working, to arrange the resistance of the primary circuit as high as practicable in order to prevent the rapid running down of the cells. One of the best of this convenient form of cell is that known as the "*British*," which is depicted in Fig. 120. The construction of this cell will not be without interest. In Fig. 121 a section of this cell is shown. It will be seen that the zinc element takes the form of a containing vessel *F*, and so answers a double

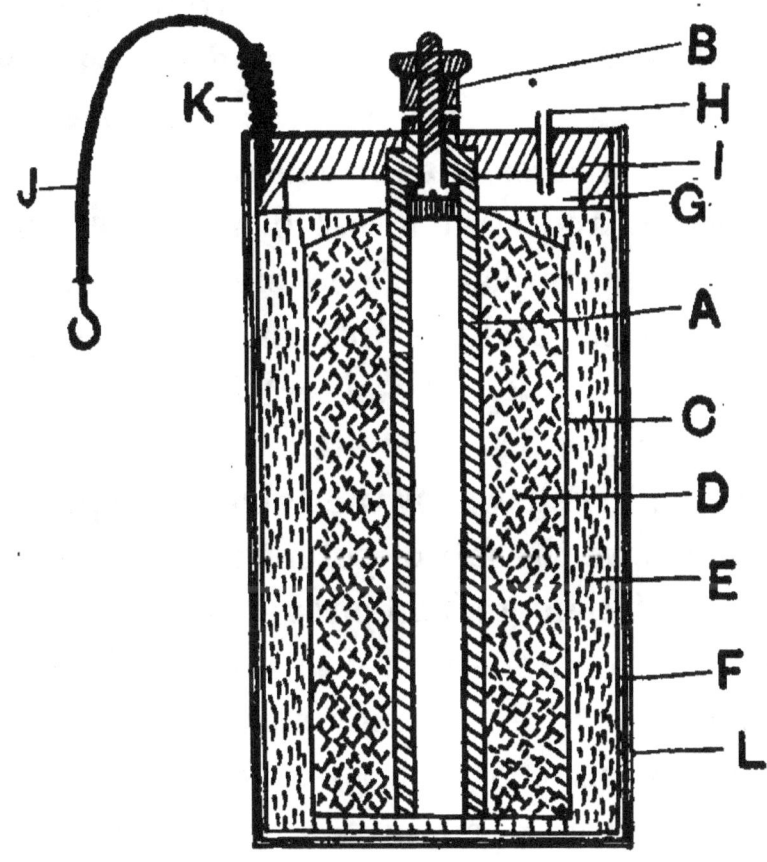

Fig. 121.

purpose. Centrally within the zinc vessel and furnished with a terminal *B*, stands the carbon *A*, which is surrounded with a black paste *D*, composed of about equal parts of powdered manganese dioxide and powdered carbon moistened with sal ammoniac solution and glycerine and enveloped in a thin porous diaphragm *C*. The space

Y

between this black paste and the zinc is filled in with a white paste *E*, consisting of a mixture of 2 parts of sal ammoniac, 1 part of chloride of magnesium, 1 part of glycerine, 1 part of flour, 6½ parts of plaster of Paris, and 6 parts of water. Above these pastes is placed a collecting chamber *G*, which allows the ammonia and other gases produced during the working of the cell to escape by means of the exit tube *H*. The zinc has two wires soldered to it, *J* and *K*. The latter is wound around the base of the former, and is provided for use in case that the former should get broken. The cell is furnished with an insulating millboard case *L*, and is then sealed with a bituminous sealing *I*, when it is complete. The electro-motive force of this cell is 1·5 volts, and the internal resistance of the largest size (8″ high × 3¼″ diameter) is 0·10 ohm, and the smallest (4″ high × 1⅜″ diameter) is 0·75 ohm. The constancy of the cell is shown by the table on page 323, in which the curve shows the fall of potential of a "*British*" cell (7″ high × 2⅝″ diameter) on circuit through 15 ohms external resistance for a period of 50 days, the current only being opened for a few seconds at intervals of a day to enable readings on the voltmeter to be taken. The · current on short circuit given by the cell before the test was 7 amps, and on the completion of the test was 0·20 amp. It is remarkable that the potential had only fallen to 1 volt on the tenth day.

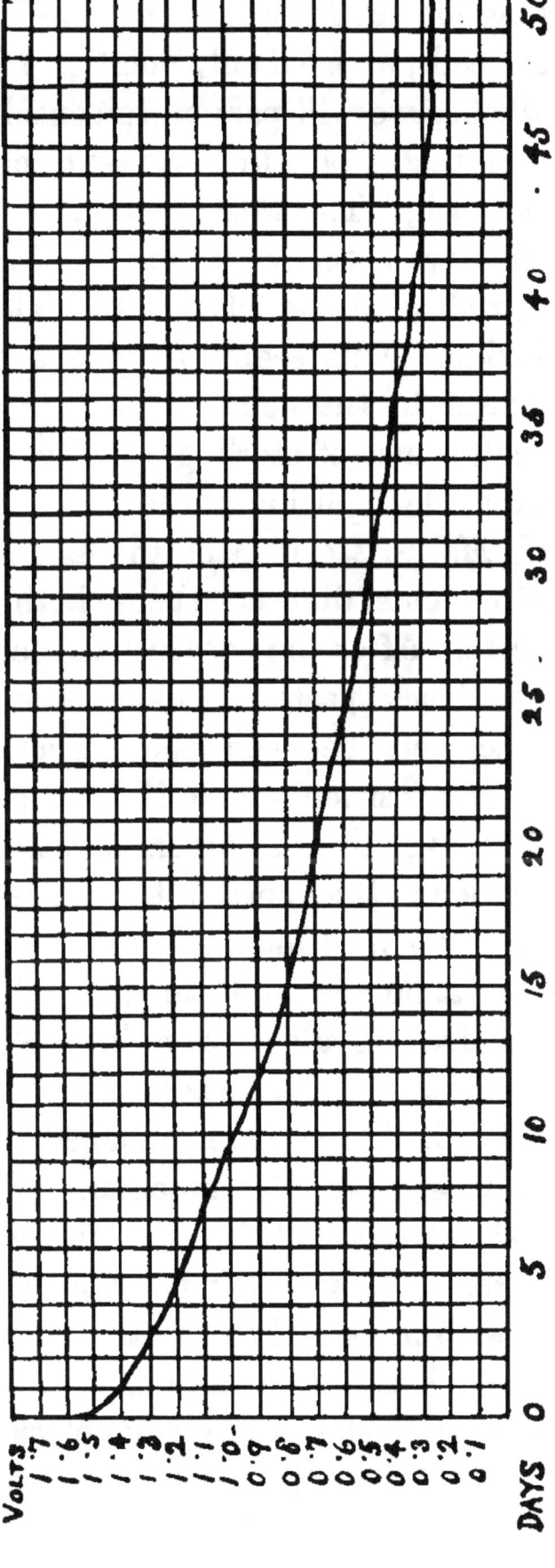

§ 254. Having gone so fully into the construction of the best forms of dry cells, we need only mention a few of similar constitution, whence we will pass on to consider those, the exciting fluids or pastes of which present marked differences.   *Dawson's* cell, " The Isle of Man," is a very good cell.   Its peculiarity is the very considerable amount of ferric oxide, mixed with the black paste, along with the manganese dioxide.   This seems to give a very marked recuperative power to the cell.

The *Meyra* cell is also of sterling quality, and is largely used by autocarists for ignition purposes.

§ 255. *C. H. Mehner's* cell is of the same class as the fore-mentioned, the chief difference being in the composition of the white paste.   After the carbon has been surrounded with the usual black paste of manganese dioxide and carbon, the space left between the carbon and zinc is filled in with a composition which Mehner calls " mineral gelatine."   To make this, a supersaturated solution of magnesium chloride is thickened to the desired consistency by the addition of calcined magnesia; and to this mixture the chlorides of ammonia and calcium, which serve as excitants, are added.   Made known in 1889.

§ 256. In *Wilson's* cell the carbon central rod takes the form of a grooved hollow cylinder, plugged at the bottom and filled with coarse granular manganese dioxide.   The usual black paste surrounds this.   The zinc cylinder, which is perforated, stands in an outer jar, and is packed both inside and out with moss soaked in sal ammoniac solution. The cell, as usual, is sealed on the top with pitch.

§ 257. The *Meserole* cell has the usual black paste around

the carbon, but the white paste consists of calcium hydrate (slaked lime) 1 part, white arsenic 1 part, glucose $\frac{1}{2}$ part, dextrine $\frac{1}{2}$ part.   These are to be intimately mixed while dry, and finally worked up into a paste with a fluid containing equal parts of saturated solution of sal ammoniac and common salt, to which is added $\frac{1}{10}$ part by volume of solution of bichloride of mercury and an equal volume of hydrochloric acid.

§ 258. *Zerda Bayon*, of Bogota, Colombia, in 1891 devised a flat form of dry cell.   The elements are zinc and carbon plates, separated by a piece of flannel.   Surrounding the carbon we have the usual black paste; on the zinc side of the flannel, we have vegetable nitrogenous matter (pea or bean flour) soaked in 20 per cent. solution of sodium chloride (common salt) and alum.   The whole, with the exception of the terminals, is enclosed in a case of india-rubber, held in place by a couple of rubber bands.

§ 259. The "*Zodiac*" cell (1896) is almost precisely similar to the Gassner, the chief difference being the large proportion of manganese dioxide used in the black paste (about $\frac{1}{3}$ of the whole, instead of $\frac{1}{8}$).   To keep the paste moist, as much as $\frac{1}{2}$ oz. of glycerine is used in each paste, in a cell $7'' \times 3''$.   E.M.F. said to be 1·62 volts, but this is hardly likely; the resistance is put down as 0·21 ohm.

§ 260. *Bryan* (1895) devised a dry cell, specially with a view to use in electric belts for medical purposes.   A curved zinc case, similar to that of a pocket-accumulator in shape, containing a smaller curved carbon plate, not touching the bottom of the zinc case, is filled nearly to the top with a mixture of sulphate of iron 6 parts, bisulphate of mercury 1 part, plaster of Paris 5 parts, water and glycerine sufficient to make a stiff paste. The solid constituents are all by measure. After the mixture has been well rammed in, the cell is sealed with paraffin-wax.

§ 261. *Rénault* (1895) appears to have been the first to employ gelatinous silex in the production of the pastes for dry cells. Gelatinous silex is prepared from sand, or powdered flint, by melting them with three to four times their weight of potassic carbonate, dissolving the fused mass in water, and then adding hydrochloric acid, which separates the silica as a jelly. A graphite cylindrical pot, similar to that used for the Judson cell (q. v.), forms the outer recipient. At the bottom of this cell, reaching up about $\frac{1}{4}$ the total height, is a mixture of equal parts gelatinous silex and chromic acid. A porous earthenware disc covers this mixture and keeps it in its place. A central spiral of stout zinc reaches from this disc to the top of the cell. The remaining space is filled in with gelatinous silica alone. E.M.F. 1·9 volts; resistance very low.

§ 262. *Bowlay* introduced in 1890 a partially dry cell consisting in a porous pot divided into two halves by a porous partition; on one side is placed a zinc plate packed in with flour of sulphur moistened with a solution of common salt. The other partition contains the carbon plate, and is filled in with either copper sulphate or potassium nitrate. The porous cell is then sealed over except for two small vent holes and the spots where the terminals protrude. The entire porous pot stands in an outer cell containing a saturated solution of sulphate of copper, with which it is fitted to within an inch of the brim. The whole is then sealed over with pitch, the terminals only protruding. This cell seems not to have found any admirers.

§ 263. A rather different cell is that of *Slater* (1882). The peculiarity lies in the employment of caustic soda or potash in the exciting paste. Sawdust, or any other form of cellulose, is boiled with a saturated solution of caustic soda, to form a pulpy mass; this forms the inner coating to the zinc. The mass which is packed between this and the central carbon consists in a mixture of quicklime slacked with bichromate of potash solution, to which is then added nitric acid and sufficient of the pulpy sawdust, that has been treated with soda, to make a consistent paste. E.M.F. not stated. Can be recharged, when exhausted, by a dynamo.

§ 264. *The Hydra "Semi-dry cell."*—This battery has been in successful and growing use in Germany and other countries for the past four years or more, and has largely supplanted many other kinds of batteries. It is the invention of Prof. Paul Schmidt, Carl Koenig, and Robert Krayn, and combines the best proportions and balance of materials calculated to produce the highest efficiency in the steadiness and volume of current under the most exacting circumstances. It received the highest award and gold medal for batteries at the Paris Exposition.

Referring to Fig. 122, which is an illustration of the section of single cell and beginning at the centre, *A* is a cylinder of zinc closed at the bottom end and open at the top, holding in its interior a special liquid electrolyte, and is also connected by an insulated wire *F* to the outer thick

zinc cylinder *E*.   The zinc cylinder *A*, it will be observed,
is placed within the carbon cup *C*, with an absorbent paste
depolarizer *B*, between the two surfaces, the cup having a
terminal at its upper end and its lower open end sealed up

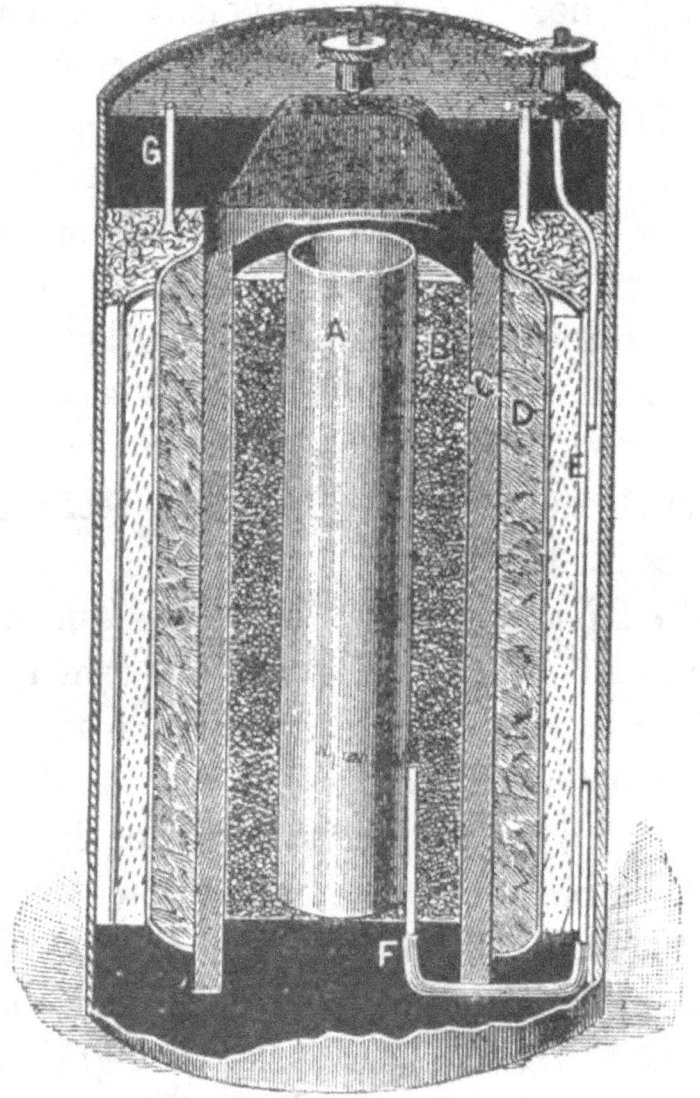

Fig. 122.

with pitch through which wire *F* passes.   Outside the
carbon cylinder is a dry-pressed generating depolarizer, *D*,
inclosed in a network of linen, which is encircled by the
outer zinc *E*, from which the other terminal rises to the

top of the cell. Over all the elements at the top is saw-dust and absorbent cotton, and from this rise vent-tubes *G*, which pass through the asphalte top. The whole is inclosed in a thin outer metal casing, insulated from the outer zinc cylinder *E*, by which the battery is protected from dampness and other injury. A small chamber in the carbon cylinder *C*, at the top just above the inner zinc *A*, allows the electrolyte in latter to overflow occasionally and keep the absorbent paste *B* slightly moistened. The moisture thus absorbed passes from *B* through the porous carbon *C* to the dry absorbent *D*, thereby maintaining it in proper condition for the generation of a current when the terminals of the battery are connected. The perfectly dry condition of the generating depolarizer *D* prevents any local internal action when the cell is not in use. The double surface of zinc provided by having an inner and outer zinc cylinder, as well as the depolarizing material on the inner and outer side of the carbon cylinder, give the battery a remarkably constant voltage, low internal resistance, and high capacity, with the added advantage that the most electrolyte is brought to the relief of the decomposing paste, keeping it permanently humid, giving the cell excellent power of recuperation after long use. The larger-sized single cell has an initial current of 22 to 30 ampères at $1\frac{1}{2}$ volts.

Although the composition of the electrolyte is not stated, yet from the E.M.F. and general behaviour of the cell, there can be little doubt but that it consists chiefly in an aqueous solution of chlorides; probably of zinc, of ammonium, and of calcium.

CHAPTER VIII

ON THE CONSTRUCTION OF THE MORE USEFUL CELLS, AND OF THEIR PARTS

§ 265. IT is not our intention to enter very deeply here into the manufacture of batteries, or their parts; but as in many instances it is convenient to be able to mount or repair a cell, without having recourse to the dealer, we shall give a few hints, which may prove serviceable.

§ 266. The outer containing vessels should be in glass, or well-glazed earthenware. Of these latter, few are so good as the jars in which table-salt is put up (by Weston and Westall), or the 3 lb. stone jam-jars of Sidney Ord. These are about 6″ high by 4″ in outside diameter, and consequently will contain a fair amount of exciting fluid, besides the elements and the porous cell. For larger cells, and especially for such as have to contain sulphate of copper, or other salts which attack the glaze, the 2 lb. and 4 lb. French plum-jars are excellent, and can be had cheaply. After having been cleaned, by soaking in water in which washing-soda has been dissolved, they should be well rinsed and dried. If they are to be used with any salts that have a tendency to *creep* (such as sulphate of sulphate of copper, ammonium chloride, etc.), they should then be immersed, mouth downwards, in melted paraffin wax, for a depth of about 2″. The mouth of the jar must be slightly tilted, otherwise the contained atmospheric air prevents the paraffin from entering the jar to any depth. If ebonite were plastic, and admitted of being easily cemented together, it would be invaluable to the amateur for making cells ; but the process of making ebonite articles from masticated rubber mixed with sulphur and compressed into heated moulds, is not one easily mastered by the tyro. Very excellent cells can be built up of sheets of brown-paper rolled round any pattern and fastened together with flour paste. Such cells should be at least $\frac{1}{8}$″ thick for sizes not exceeding 6″ × 3″ + 3″. When thoroughly dry, they must be boiled in melted paraffin wax (at about 135° F.) until bubbles cease to appear. The cells are then withdrawn from the paraffin, and allowed to drain and cool.

Another excellent material with which to make cells is *celluloid*. This can be procured in sheets about $\frac{1}{8}$″ thick, nearly as transparent as glass, extremely strong, and easily cut or joined. The natural cement is a solution of itself in amyl acetate ; in the proportion of one part celluloid to fifty parts amyl acetate. Celluloid is a composition of gun-cotton with camphor and castor-oil, and hence is highly inflammable : this perhaps is its only defect.

§ 267. Where porous divisions or cells are required, it will generally be found much cheaper to buy than to make. For very small experimental work, the bowl of an ordinary " clay " pipe, the stem being broken off short, and the hole plugged with paraffin wax, answers admir-

ably.   A new No. 60 flower-pot, with the draining-hole
plugged with a cork, will do good service.   When it is a
question of using either flat cells, or simply divisions, it
will be found more satisfactory in every way to procure
them from the potteries.   But where it is impossible to
have recourse to these, temporary substitutes may be
found in cells or slabs cast in plaster of Paris, and allowed
to dry thoroughly before use.   Animal membranes, such
as bladder, parchment, an ox-gullet (tied below), have
been used as porous septa.

§ 268. As we have already seen, most batteries depend
on zinc as their positive element.   For this reason a few
remarks as to the mode of cutting and amalgamating zinc
may not be out of place here.   Very thin sheets, scraps
for melting or for making up composite cells, can be cut
with a large pair of scissors, or with the tinman's snips.
Stouter sheets, that is, anything between $\frac{1}{32}''$ and $\frac{3}{16}''$ in
thickness, must be cut either with a heavy pair of shears,
with a circular or other saw, by chiselling a line along and
then bending, to decide fracture, or, lastly, by filing or
scratching a furrow along the desired line of separation,
pouring mercury into this furrow.   Very soon the mercury
is absorbed by the freshly-scratched line in the metal
rendering it extremely brittle; when a smart blow will
decide the fracture along the thin line.   All cells in which
acids are used will work more economically and with less
local action, if the zinc be amalgamated.   If salts or
alkalies are employed as excitants, the good effects of
amalgamation are not so marked; in fact, it is very
doubtful whether any benefit is obtained; and although

it is customary to amalgamate the zincs of the Leclanché and similar slow-acting cells, the good results are very problematical. There are two modes of amalgamating zincs, (*a*) by the direct application of mercury to the zinc; (*b*) by the action of a solution of a mercury salt on the zinc, under the influence of which mercury is liberated, which then combines with the subjacent zinc. The former method gives the better result, as the mercury penetrates much deeper than is the case with the latter process. In the first method, the zinc is first scrubbed with a hard brush and common washing-soda and water, then rinsed, and laid in a flat dish containing clean water to which a few drops of hydrochloric acid have been added. A few globules of mercury are then poured on the surface of the zinc and briskly rubbed over both surfaces with a pledget of tow. The zincs are then reared up on edge on another dish, when the superfluous mercury drains off, and may be collected for use on another occasion. By the second method, the zincs, after having been cleaned from grease as before, by scrubbing with a solution of washing-soda and rinsing, are rubbed over with a pledget of tow dipped in a solution of nitrate of mercury in water; or into a paste of mercury sulphate and water. In either case, the rubbing must be continued, with the addition of mercury or of the mercurial salt, until the whole surface of the zinc becomes as resplendent as a mirror.

§ 269. Zinc rods may be cast in *dry* plaster of Paris moulds; but these may now be bought so cheaply, that it is hardly worth while attempting to make them, unless indeed some peculiar form is desired. Care must be taken that the mould be *absolutely dry*, before attempting to cast therein, since, as the zinc must be nearly red-hot to enable it to flow freely, it would give rise to a serious explosion if the mould were at all damp.

§ 270. Another element which enters into the construction of most cells is *carbon*, in the form of graphite. Plates, blocks, rods, and cylinders of this material, made by baking a paste of powdered retort-scurf mixed with sugar and compressed into shape in moulds previous to baking, can be purchased at very moderate rates. But with a few notable exceptions, these moulded plates, etc., are neither so hard, nor of such high conductivity as plates cut from the hard carbon deposited in the inside of the gas retorts, and generally known as " gas-carbon " or " retort-scurf." Where small pieces or peculiar shapes are required, and more especially when high conductivity is a desideratum, it is worth while to procure a large piece of retort-scurf from the nearest gas-works, and cut it to the desired shape. An old saw, fed with silver-sand and water, aided by patience and elbow-grease, will enable the enthusiastic operator to cut out the larger pieces from the curved cake that comes out of the retorts : a disc of iron, with its edge " upset " by hammering, mounted on a spindle, and set rotating between the centres of a lathe, will cut the blocks thus produced into thin sheets, small blocks, or irregular forms as required ; lastly, very thin sheets can be obtained from the larger ones, by sprinkling some fine silver-sand and water over the surface of a perfectly flat stone slab, and grinding the carbon by rubbing on the slab with a circular motion. This is dirty work, but very efficient; and with care, sheets of graphite about $\frac{1}{10}''$ to $\frac{3}{32}''$ can be obtained by this means. For most purposes carbon is the best negative element for a battery, better even than platinum, owing to its roughness presenting such a large surface: its chief defect being its porosity, which allows the solutions into which it may be immersed to creep up and attack the terminals to which it is attached. There are several ways of lessening this evil. The first, and most efficient, is that of placing a piece of thin platinum foil between the carbon and the jaws of the clamp or terminal. But with the present price of platinum this is an expensive method. The second mode is to cast a leaden head to the plate. Lead is but little acted on by dilute sulphuric or chromic acids, and hardly at all by sal ammoniac; so that this remedy is fairly satisfactory, if the precaution be taken of plunging the carbons in boiling water *nearly* up to the lead capping, as soon as the battery or cell has been made use of; after which the carbon should be allowed to dry.

§ 271. Sometimes it is convenient to make a soldered joint between a carbon plate and a piece of metal; as, for instance, in the case of the two carbons in a "bottle bichromate" (see § 61), in which the two carbon plates are joined together, so as to form but one element, by a perforated ⊓ shaped strap of copper or brass, which is itself screwed to the under surface of the cover of the bottle, and to which the positive terminal makes connection. Since carbon of itself cannot "take" solder, it is evident that the surface must in some manner be metalized. A very simple mode of effecting this is by depositing electrolytically a coating of *copper* on that part of the carbon to which it is desired to apply the solder. The secret of success is to perform the operation quickly, so that the copper solution may not have time to soak into the pores of the carbon. The following mode of procedure gives excellent results. Having cleaned thoroughly and "tinned" with the soldering-iron, the piece of brass or copper to which it is intended to solder the carbon, and having the solder, soldering-fluid, and soldering-iron all at hand, the latter cleaned and ready to be heated at a moment's notice, the carbon plate, which should be *quite* dry, is cleaned at the portion intended to be coppered, by rubbing over with another flat piece of carbon, and wiping with a perfectly clean rag. The opposite extremity is tightly bound round with a few turns of No. 16 copper wire, which is then tightly twisted, to prevent uncoiling, a long end being left free for attachment to the negative pole of a quart size bichromate cell. To the positive pole of the said cell, a piece of sheet copper, about the size of the carbon, is attached by means of a few inches of copper wire. Both plates are then plunged into a glass vessel containing a saturated solution of sulphate of copper in water, to which a few drops of sulphuric acid have been added (about ten drops to the pint). The copper plate may be entirely immersed in this solution; but the carbon plate should not be plunged farther in than the surface over which it is desired to deposit the copper. If the battery is working properly, copper is quickly deposited on the carbon, and in from fifteen to twenty minutes will

have formed a coating of sufficient thickness to enable
the carbon to hold the solder. When the operator judges
the coating is thick enough, he will disconnect the carbon
wire from the cell, and immediately hold the carbon plate
under a running stream of water. Then with the hot and
freshly tinned soldering-iron in his right hand, he will
apply to the coppered portion with a feather held in his
left, a film of soldering-fluid, followed by the soldering-
iron, applying more solder, if necessary, so as to produce
a fairly thick coating of solder on the copper. The piece
of brass or copper is then pressed against this surface,
and heat applied either with another and hotter soldering-
iron, or by the flame of a Bunsen burner or spirit-lamp,
until the solder on the carbon and the brass surfaces flow
together. When *quite* cold, the join will be found to be
perfect, and should be painted all round with good Bruns-
wick black, to prevent any acid coming into contact with
the brass piece, or with the deposited copper; as this
would soon corrode the metals, and thus rapidly spoil the
connections.

§ 272. Connections to zinc plates or rods can be made
in several ways. Either the plate may have a hole drilled
along its upper edge, which hole is then tapped with a
female screw, so as to receive the shank of the terminal,
or other screwed wire; or the plate may be gripped
between the jaws of a metal clamp; or in the case of a
rod, a wire may be cast in the top extremity, or if the
rods are already made, without wires attached, a fine slit
cut in the top with a fine saw, the wire hammered flat at
its extremity to fit the slit, forced in, and then soldered

z

round the top, in the usual way, with spirits of salt only, as flux.

§ 273. In mixing solutions, care must be taken in order to avoid dangerous accidents. In all cases in which strong stilphuric acid has to be employed in conjunction with water, the *acid* must be gradually added to the water, with constant stirring; and *not* the water to the acid. As the mixture *heats*, the mixing should be done in a glazed stoneware vessel, and not in glass. If by any chance the clothes or hands get splashed by the acid solution, immediate application of a strong solution of washing-soda, or of liquor ammoniæ, will neutralize the acid and minimize its corrosive action. Excitants containing caustic potash or caustic soda do not "burn" so severely as sulphuric or nitric acid; but their effect is more insidious, giving rise to disintegration of the flesh, and sometimes producing bad sores. At first the skin feels simply comfortably smooth and soapy; but the alkali soon eats its way into the flesh, when much pain is felt. It is advisable therefore, after touching alkaline solutions, to wash the hands, etc., in plenty of vinegar, and afterwards in water. This treatment will obviate all inconveniences.

§ 274. We now proceed to describe the mode of constructing an efficient form of two-fluid cell, which can be used either with chromic acid as the depolarizer (as in the "Boron" cell) or with nitric acid, like a Grove or Bunsen cell. We have chosen this form, as its construction illustrates the mode of dealing with cells of square or other uncommon shapes; containing vessels for which are not found as stock articles at the dealers', as are cylindrical vessels. We begin by making or procuring a rectangular porous cell, about $5\frac{3}{8}''$ high, by $4\frac{1}{2}''$ wide, and $1''$ deep from back to front. We then make a square box either of celluloid, gutta-percha, or wood, the inside dimensions being $5''$ high by $4\frac{3}{4}''$ in the sides. If made of wood, the component pieces, after being planed up and accurately fitted, should be cemented together to form the box, with thick shellac varnish, and the edges and bottom tightly drawn together by means of $1''$ iron screws, about $\frac{1}{8}''$ in diameter, with flat heads, driven in flush with the wood; the thickness of the wood from which this box is to be made should be $\frac{3}{8}$ of an inch when planed up. After the box has been constructed, it should be boiled in melted paraffin wax, until bubbles no longer rise to the surface. This renders the wood impermeable by acids or water. Too great heat should not be applied, only sufficient to keep the paraffin wax melted, otherwise the wood will warp. The box should be withdrawn from the paraffin, and allowed to drain. It is advisable not to drain the *inside* too closely, as it is advantageous that there should be a pretty thick coating of paraffin left in the interior. If *celluloid* be chosen as the material for the outer vessel, the five pieces, after having been cut accurately to size, are fastened together by painting over the surfaces to be joined, with a solution of clear celluloid in *amyl acetate*, of the consistency of ordinary gum-water. The parts are then clamped tightly together, and allowed to stand undisturbed for twenty-four hours. This forms a very strong and homogeneous joint. Celluloid is an excellent material for cells: it is not attacked by any acid, is almost as transparent as glass, and is an insulator of the highest order. Its only defect is its *inflammability*, due to the fact of its being made chiefly of xyloidin (a form of gun-cotton). If gutta-percha be selected, the component pieces can be joined together by heating along the edges to be attached, with a fairly hot soldering-iron, and then pressing the portions into close contact.

§ 275. The outer cell being made in any of the modes above described, the porous cell is placed in it, in the middle, so as to divide it into two equal portions. This is retained in position by two strips of wood previously boiled in paraffin wax, each $\frac{3}{8}''$ square, and sufficiently long to overlap the edges of the box, by about $\frac{1}{2}''$ on each side. One of these strips is placed on each side of the porous cell, *not actually touching*, so that the porous cell may be withdrawn if required, and the two strips joined together on their under surfaces, when they project over the box by two shorter pieces of wood, held by $\frac{3}{4}''$ flat-headed screws. A zinc plate is now cut to fit easily in the porous cell, about 6″ long, 4″ wide, and $\frac{1}{8}''$ to $\frac{3}{16}''$ thick. Across the centre of the top edge of this zinc plate, about $\frac{3}{8}''$, is drilled or punched a $\frac{1}{8}''$ hole, and a copper strip 4″ long, $\frac{5}{8}''$ wide, and about $\frac{1}{30}''$ thick is riveted to it, at right angles, by means of a copper rivet. This copper strip serves as one terminal or connection, and to it may be attached, if desired, a binding screw; the zinc plate must now be amalgamated, as described at § 268. (All work to be done to zinc plates, etc., must be performed *before* amalgamation, as this process renders it very brittle.) A pair of carbons should now be cut, preferably of gas-carbon, 5″ long, 4¼″ wide, and about ⅜″ thick. These are to stand, one each side of the cell, on the outside of the little wooden frame. These two carbons are to be connected together across the top by a saddle-shaped strap of rather stout sheet brass, which should completely clear the two wooden strips and the porous cell. This strap, which should be 1″ wide and the shape of a ⌒, is best attached to the carbons, by drilling a hole in each limb, and corresponding holes having been drilled across the top edge of each carbon plate, a small flat-headed screw, fitted with a nut, is passed through each pair of holes, and the nuts then tightened up, until the carbon is held quite rigidly. A terminal can then be soldered to the centre of the ⌒ strap. This mode of connection permits of the carbon being easily detached from its connections, in case these become corroded, therefore requiring cleaning. It is needless to remark that should this cell be employed with *nitric* acid (as a Grove or a Bunsen cell) it will be better to place the carbon in the porous receptacle and two zincs on the outside, thus reversing the position described above.

§ 276. We close this chapter with a description of the mode of constructing a cell with trough for containing crystals, which can be used for a Daniell, or similar type of battery. A sheet of No. 24-gauge copper, 6″ wide and 19″ long, is bent round a form, so as to produce a cylinder 6″ in diameter, and the two edges bent over in opposite directions to a depth of $\frac{3}{16}$ of an inch, thus ⅃ ⌐, so as to hook into each other. These edges are then lapped over and beaten down closely with a mallet on a stout iron rod, after which the suture is neatly soldered down on the *outside*. A copper disc is now cut from a similar piece of sheet, about 6¼″ in diameter, to form the bottom of the cell. The bottom edge of the cylinder should be slightly splayed outwards for about ⅛″ all round, by careful hammering against a sharp edge; the cylinder being then placed on the disc, which should previously have had *its* edge bent upwards to the same extent. The bent edge of this latter should then be turned right over the splayed edge of the cylinder, and then hammered flat all round; after which, to ensure its being watertight, a little solder should be run round the edge thus turned over. Taking a medium-sized binding screw, we file away one side of the shank, so as to leave it flat, instead of round, and we solder this shank to the outside of the upper edge of our copper vessel. We now require a porous cell (cylindrical) 7″ in height by 2″ in diameter, to which we adapt a loosely-fitting wooden cover, having a slightly projecting flange. To prevent this swelling by damp, it should be boiled in melted paraffin wax. A small central hole is put through this, to admit the passage of a fairly stout copper wire, as through this hole will be passed the wire of an ordinary No. 2 size Leclanché zinc rod. The projecting portion of this wire may be soldered to the shank of a second terminal outside the wooden cover. When the zinc wants renewing, it is then an easy matter to unsolder the terminal by touching the junction with a hot soldering-iron. The cell proper is now complete : but, as in cells of the Daniell type, it is necessary to keep a supply of sulphate of copper (or other) crystal at the surface of the depolarizing solution; in order to maintain its strength, it is advisable to make a kind of perforated basket or tray which dips a little way into the liquid from above, wherein is placed a supply of the crystals needed. To make this, the operator takes a piece of copper sheet, of the same gauge as before, about 4″ wide and 8½″ long. He makes a line ¾″ from the edge along its length, and with a pair of shears cuts along this line for about 1″ in from each extremity. He then cuts down perpendicularly to this line until he meets the cut, and thus removes the two corner pieces. At about ¾″ from this last cut, and parallel to it, he cuts down till he reaches the first line, performing this operation at both ends. He now bends back the ¾″ pieces, so as to enable him to cut along the line from one bent piece to the other. By this means he produces a sheet 8½″ long, 3¼″ wide, with two ¾″ lugs projecting along one edge. Bending this sheet so as to make it fit accurately to the curvature of the inside of the copper vessel previously made, he turns the lugs over the outside of the copper vessel, so as to hold the semi-circular strip firmly in position, yet allow of its removal if required. Taking another piece of the same sheet copper, he cuts out a figure 6″ long along one edge with two sides at right angles to this edge, 3″ long from whence it terminates in a semi-circle of 3″ radius. With any sharp point he makes a number of perforations all over the surface of this piece; and then cuts a little nick in each of the 3″ sides, also in the edge of the semi-circular portion. These pieces he bends outwardly with a pair of pliers, so that they project out as lugs. He then bends the straight portion to make a right angle with the semi-circular part. Placing this against the 8½″ curved piece hanging on the vessel, with the semi-circular portion downwards against the curvature of the said 8½″ piece, and the straight portions flush with edge top thereof, he marks with a scratch the places where the little lugs reach. He makes a perforation at each of these points, pushes the little lugs through, and bends them tightly over, by this means forming a little perforated basket, *without using any solder.* This is essential, as any solder would be acted on electrolitically, and would speedily be corroded and destroyed. The cell is now completed, and the porous cell with its contained zinc is to be placed in the copper vessel, in the space left between the movable shelf or tray, and the other side of the copper vessel. To charge this, dilute sulphuric acid (1 to 20 of water) may be placed in the zinc compartment; while in the copper vessel, a saturated solution of sulphate of copper, acidulated with about 1 per cent. of sulphuric, is employed. Crystals of sulphate of copper should be placed in the basket. By osmose, the sulphate of copper solution gradually enters the porous cell *when the cell is not acting,* thus spoiling the zinc; this effect does not take place if the cell is generating current; so that it is advisable, if the cell is to be put aside for any length of time, to leave it connected up, with a very high resistance in circuit. By this means a trifling current is set up, entailing but little waste, which effectually protects the zinc from deposition of copper on its surface.

CHAPTER IX

ACCUMULATORS : CONSTRUCTIONAL DETAILS

§ 277. At Chapter V. we studied the secondary cell or "accumulator" from a theoretical point of view ; here we propose to give a brief *résumé* of the modifications which have been introduced in order to render the construction easier, or the output greater.

In 1801 Gautherot showed that platinum or silver wires, when used for the electrolysis of saline solutions, gave secondary currents on the cessation of the primary current. Ritter, very shortly after (1803), showed that piles formed simply of discs of copper separated by rather smaller discs of card or flannel previously moistened with a solution of sal ammoniac in water, although incapable of setting up current on their own account, gave powerful discharges after being connected for some time with a battery giving sufficient E.M.F. to effect the decomposition of the water. After this, Volta, Davy, Becquerel, Marianini and others found that many substances, such as gold, silver, platinum, etc., were capable, after being used for electrolyzing certain solutions, of giving transitory secondary currents when the primary current was interrupted. In 1842 Grove constructed his gas battery, of which we have given a description at § 33. With a series consisting of 50 of these cells, previously charged by means of a suitable primary battery, he was able to produce an arc-light, to effect chemical decomposition, and, in fact, to perform all the operations for which a primary battery is usually employed. In his *Experimental Researches* Michael Faraday describes an experiment, wherein the electrolysis of a solution of acetate of lead furnished him with a deposit of metallic lead at the negative pole, and of lead peroxide at the positive pole of his electrolytic bath; and he points out (a fact which was afterwards applied to practical use) the high conductivity of the lead peroxide and its readiness to part with its oxygen. Wheatstone, De la Rue, Niaudet and others seem to have shared this knowledge that lead peroxide, or even minium (red lead), which is a compound of lead monoxide with lead peroxide, having the formula of $2PbO + PbO_2$, might be employed by virtue of its conductivity and the readiness with which it gives up its oxygen, as a powerful depolarizer in primary batteries.

To Gaston Planté belongs the merit of having constructed the first practical accumulator or storage cell. In 1859 he described an arrangement consisting of two similar sheets of lead, about a foot wide and five or six feet in length, separated from actual contact with each other by means of strips of indiarubber or of coarse felt, the two plates then being rolled on a wooden former so as to make a compound spiral, as shown in our Fig. 123. Lugs in lead extend outwardly from the beginning of the inner and the end of the outer sheets. In order to prevent the spiral from coming undone, it is held together at its

upper extremity by a star-shaped gutta-percha clamp, and
for greater security a couple of indiarubber bands may be
sprung over the outside of the spiral. The lead plates are
now inserted in a cylindrical glass vessel of sufficient

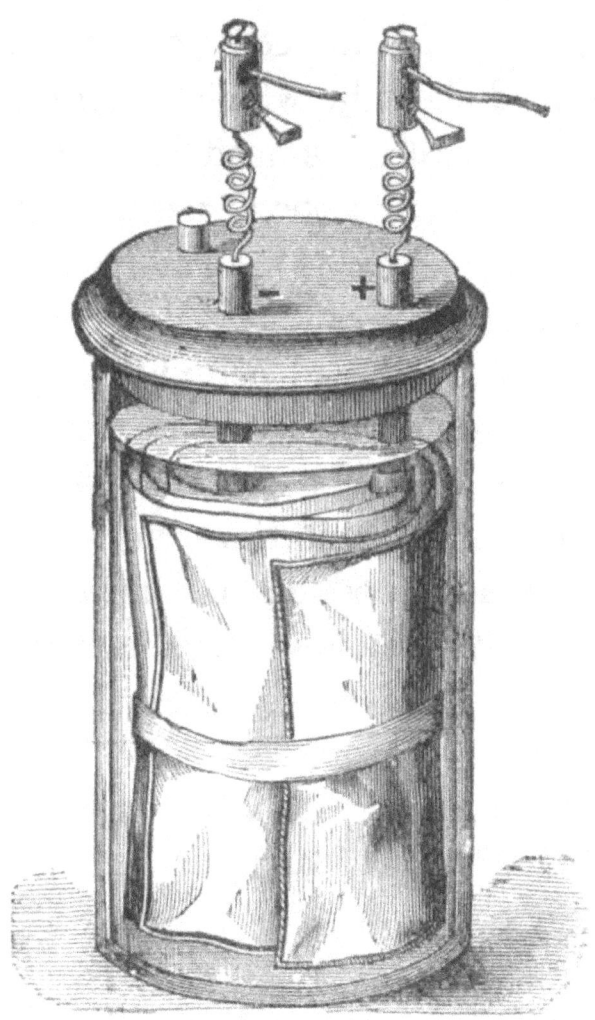

Fig. 123.

capacity to receive them along with the necessary acid;
about 14″ in height and 6″ in diameter. The glass jar
is then filled with sufficient dilute sulphuric acid, of the
strength of 1 part of pure acid to 10 parts of water, to
rise above the edges of the plates and cover them to the

depth of about $\frac{1}{2}''$. The whole is now closed with an ebonite cap, having apertures through which the leaden lugs, or terminals, can pass. On passing a current from say two Bunsen's cells (or any source capable of giving a pressure of not less than 2·5 volts) so as to cause a current to pass through the Planté cell, lead peroxide will be formed on that lead sheet by which the current enters the cell, or the anode, while the lead sheet from which the current leaves the cell, or cathode, will become coated with adherent hydrogen (see § 34). This causes the cathode plate to assume a rough granular surface, while the anode acquires a thin coating of lead peroxide. If now the contact with the charging battery be broken and the circuit between the two lead lugs be established, a powerful current passes, while at the same time the lead peroxide loses its oxygen, the hydrogen which has not escaped from the surface of the opposite plate re-combining therewith to form water, the surface of this latter plate being attacked by liberated oxygen, in proportion to the deficiency of the hydrogen lost by escaping into the atmosphere. If the Planté cell is once again charged, the anode plate, being now more spongy on its surface (owing to the reduction of the lead peroxide originally formed), takes up the oxygen more quickly and to a greater depth than it did before, while the surface of the cathode becomes reduced to the condition of spongy lead by the action of the hydrogen, and is in a condition to retain and to conduct a larger charge. For this reason, by repeated charging and discharging, the Planté cell acquires greater efficiency and capacity. This operation of reducing the cathode plate to the spongy condition and the anode to a virtually thick crust of lead peroxide is called *forming*, and the efficiency of the Planté cell depends almost entirely on the extent to which this forming is carried. As a matter of fact it has been somewhat farcically said that the Planté cell is at its best when its plates are just dropping to pieces; that is to say, when by repeated charging and discharging the entire mass of the plates has been reduced to so spongy a state that they will hardly hold together. But the "formation" of plates by battery or dynamo is a long and costly operation. The usual method employed was to charge the cell, allow it to rest, and then to reverse the direction of the current through the cell. At each reversal of the current the peroxide originally formed is reduced, first to the state of lower oxide, and finally into metallic lead in an extremely finely-divided condition by the action of the nascent hydrogen evolved at its surface during the decomposition of the electrolyte by the action of the charging current. Successive reversals had the effect of causing these oxidizing and reducing actions to penetrate more deeply into the substance of the lead, until at last almost any desired depth of active material could be obtained. It is necessary, however, to allow some days or even weeks to elapse between each reversal, in order to give time to the peroxide of lead to crystallize and become compacted together, otherwise the hydrogen evolved gets between the live metallic lead and the coating of peroxide, and causes it to scale off. It will be evident from this, that although extremely convenient in other respects and very efficient electrically, the Planté cell is not a commercial success, and the ingenuity of the inventors was soon turned in the direction of finding a means of accelerating the "forming" of the negative and positive plates. Planté himself found that by roughening the surface of the lead plates by scouring, or by immersing them in very dilute nitric acid, by applying heat during the charging process, etc., the surfaces were more quickly brought to the desired spongy condition, so that the conversion of the one surface into peroxide of lead and of the other into spongy metallic lead was more rapidly effected.

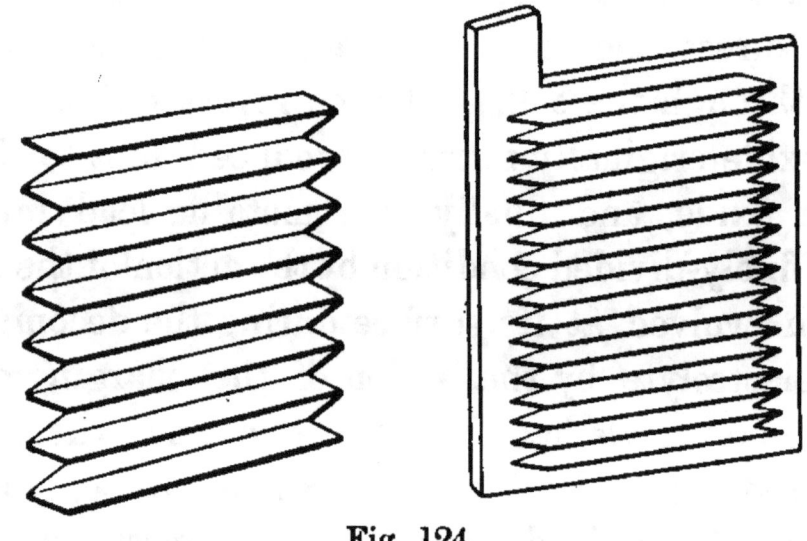

Fig. 124.

§ 278. It was soon noticed that the amount of charge which could be drawn from a given cell was largely dependent on the extent of surface which the plates presented to the action. To increase this surface M. de Meritens constructed his plates of laminæ of very thin sheet lead folded over one another like the leaves of a book or like the folds of a fan, the whole being then inserted in a strong frame of lead, to which it was soldered with an autogenic lead joint. This mode of construction furnished plates having a very large surface with a certain amount of porosity between the laminæ, so that even when the forming was effected electrolytically by battery or dynamo, much less time and less expenditure of energy were required than in the older method. We give at Fig. 124 an idea of the construction of a de Meritens plate. It must be borne in mind that in all cases where leaden plates have to be immersed in acid, any joins or solderings must be done with *lead*, and not with ordinary solder, which would be

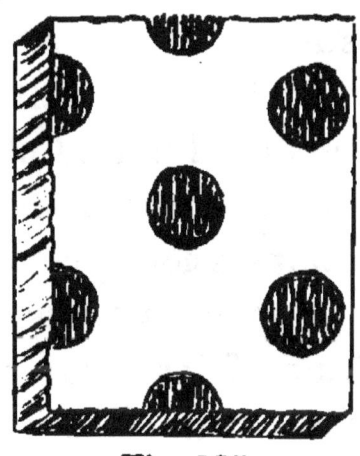

Fig. 125.

immediately attacked by the acid, thereby setting up local currents, and finally breaking down the connection between the parts. Hence the necessity of autogenic soldering. In the Kabath form, some narrow strips of lead are fluted as in the de Meritens, and then placed between two perforated sheets of lead, the whole forming one element. As this form is more obsolete, we content ourselves with giving a rough sketch of a portion of one plate only (Fig. 125). It was soon found that a charged lead peroxide plate gradually loses its electrical energy through local action, and that when the elements employed have an extensive metallic surface as compared with the amount of peroxide, this local action is necessarily large, and the cell loses its charge quickly. To increase the surface and to facilitate the " forming," many other modifications have been employed. Reynier built up his plates of lead wire ; Montaud formed his of laminated lead on which finely-divided metallic lead was deposited electrolytically from a solution of lead in caustic potash, the deposit being afterwards compressed to insure adhesion. Woodward made porous plates by stirring up melted lead along with common salt, and compressed the mixture while still plastic in moulds of the desired form, and finally dissolved out the salt by means of water. Watt of Liverpool produced a kind of lead wool by blowing steam through the molten lead and then forming this wool into sheets by compression. Simmen poured molten lead through a kind of metal colander having fine perforations, whence the kind of lead wire ran through and fell into a recipient of cold water. The sudden chilling caused the surface of the wire to break up and become rough ; this wire was then placed in suitable moulds and compressed so as to form blocks. Dujardin formed his plates from corrugated sheets or strips of lead lying close together on a leaden frame, by the electrolytical deposition of peroxide of lead from a mixture 10 parts of water, 2 parts of sulphuric acid, and 1 part of any alkaline nitrate.

§ 279. Camille Faure, seeing the great loss of time and the great expense incurred in the " formation " of plates by the methods just described, introduced his system of " pasting " the plates. Originally his plates were of the spiral form similar to the Planté, but he finally adopted, as being more convenient, flat plates. The surface of the leaden plates having been scored or otherwise roughened, are smeared over with a stiff paste made of good red lead mixed with sufficient dilute sulphuric acid (1 part strong acid sp. g. 1·845 to 1 part of water) to give it the consistency of putty. This paste is spread over the surface of the leaden plates while they are resting on a flat board by means of a wooden or bone spatula. When one side is dry the plates are turned over and the operation repeated on the other side. Sometimes, and more especially in the case of the negative plate, the paste contained, instead of red lead, a mixture of litharge (which is a protoxide of lead). In either case the passage of an electric current through a pair of such pasted plates, immersed in dilute sulphuric acid, resulted in the very rapid transformation of the red lead into lead peroxide on the positive plate, and into finely-divided or spongy metallic lead at the negative plate. This of course very considerably shortened and cheapened the operation of " forming"; but there are several drawbacks, the first being the facility with which the paste breaks away from the plates. Any little heating, or expansion, or buckling arising from a rapid discharge, causes the paste to fall away. To prevent this Faure wrapped his plates, first in parchment paper, and then in stout felt. But the presence of these bodies, apart from the facility with which they are attacked by the acid contained in the cell, militates against its efficiency, owing to the resistance they present. The ingenuity of inventors was therefore soon turned in the

direction of revising plates of such a form as should retain
the paste without any extraneous aid, and the names of
Sellon, King, Volkmar, Phillippart, Parker, Swan, Drake
and Gorham, Eickemeyer, Roberts, etc. etc., are but a few
out of about 1000 who figure as patentees of improvements
in accumulator plates between 1880 and the present date.

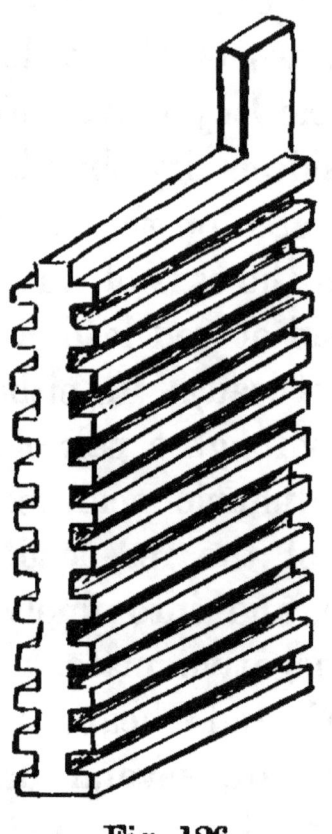

Fig. 126.

One of the first improvements consisted in punching holes
through the surface of the lead plates, which were afterwards pasted in the manner just described. Little pellets
of paste thus forced into the holes hold thereto fairly well,
especially if the surface of the lead be roughened by
previous boiling in dilute nitric acid. Epstein cast his
lead plates in the form of a fishbone with a central spine

and lateral ribs as shown at Fig. 126, and plates of this form are found to retain the paste well. The Electrical Power Storage Co., availing themselves of the patents of Messrs. Sellon and Volkmar, soon produced a plate or "grid" which was cast in lead and in which the perforations were pyramidal, that is to say, openings of about ¼″ square at the outside, diminishing to about ⅛″ in the centre of the plate. A portion of such a plate is shown

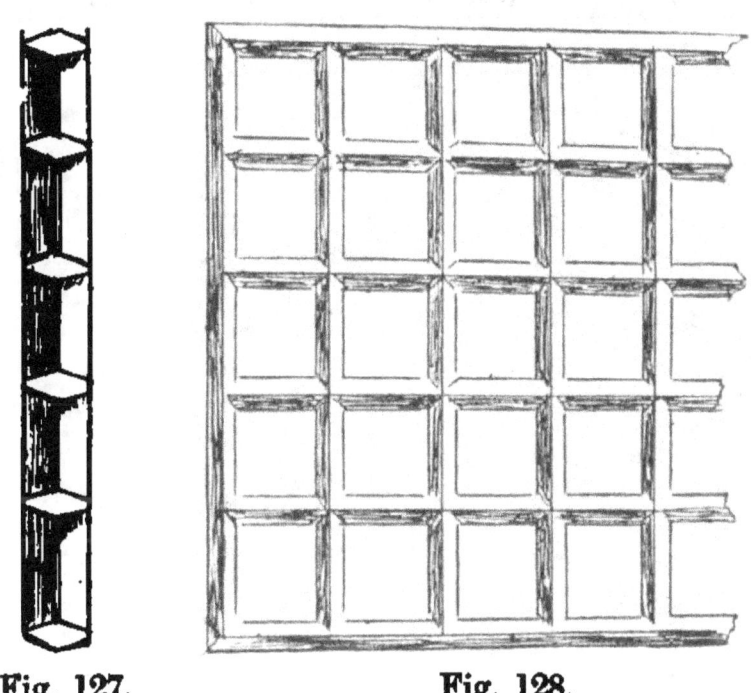

Fig. 127.   Fig. 128.

at Fig. 127 in section, and at Fig. 128 in front elevation. By this means, when the paste had once set, it could not easily disentangle itself from the meshes in the plates or grids. Many modifications in the shapes of the holes in the grids have been made. Thus Drake and Gorham, after having filled the holes with paste, compressed them by running them between rollers, which has the effect of burring over the edges of the metal and thus effectually preventing the paste from falling out. Mr. Hagen of Kalk, near Cologne, and also the brothers Jacquet of Vernon, France, make their grids in two halves, the holes or perforations being larger in the middle than on those portions which would afterwards form the two outsides. After pasting, the two plates forming the one grid are clamped, soldered, or otherwise bolted together. In the high-discharge type of the E. P. S. Co. known as " K," the positive plates have upturned ridges or ledges into which the paste imbeds itself (Figs. 129, 130).

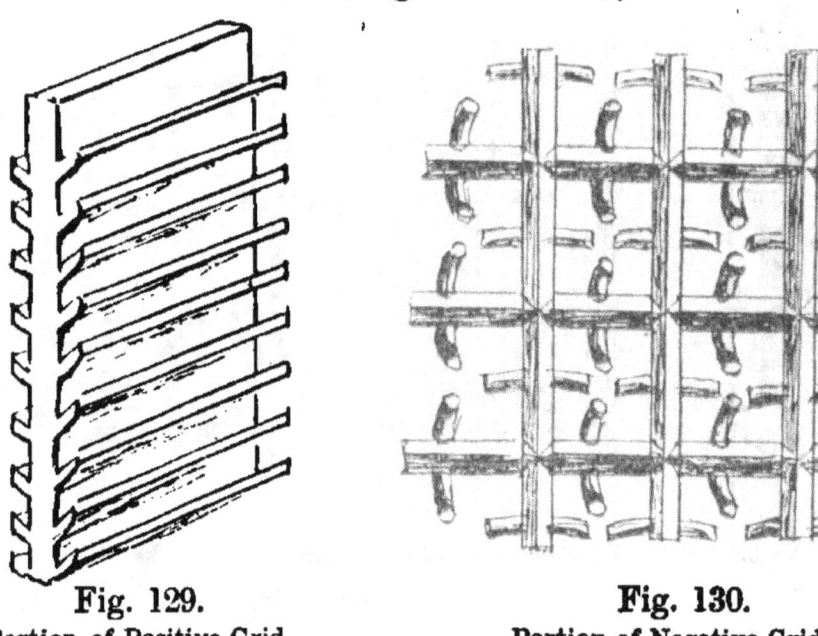

Fig. 129.
Portion of Positive Grid.

Fig. 130.
Portion of Negative Grid.

§ 280. We need only mention the names of a few inventors of modifications of accumulator plates, such as Gadot, who made his plates of double grid form, enclosing paste to the extent of 54 per cent. of the entire weight of the plates : the Pitkin and Holden " Ladder " Plate, which, as the name indicates, consists of a number of thin cast leaden troughs or trays filled with a paste and placed one over the other like the rungs of a ladder, then held together by a clamp, when the two side edges are burned together in a blowpipe flame. In order to allow space for the fluid to circulate freely between the trays or troughs, narrow strips of lead about $\frac{1}{16}''$ thick are laid between the side edges of each succeeding tray previous to burning these edges. A somewhat similar idea is carried out in the "Intercirculatory" cell introduced in America by Mr. Victor Ernst. In this an outer leaden frame retains a number of cores made of asbestos fibre, wool or hair, around which is compressed the peroxide paste or the spongy lead, these plates are then placed horizontally one over the other. In Payen's accumulator, which was the forerunner of the Chloride cell, we have a plate made of an intimate mixture of asbestos fibre and chloride of lead, which are fused together and poured into a mould. These are then reduced to the state of spongy lead by using them as cathodes against ordinary lead plates as anodes, and from these spongy lead plates the positives can be formed in the usual manner. H. Carpenter of New York produced his elements by making perforated hollow lead boxes, in which he packed cerussite (native carbonate of lead), that can be reduced by electrolysis to peroxide in the one case and to spongy lead in the other. A form not unlike Cruikshank's trough battery (see Fig. 28, § 40) was designed by Mr. Mark Bailey. It consists of a rectangular wooden box divided into separate cells by the battery plates. These plates are cast with deep square recesses reaching about half-way through the plate on each side, and these recesses are pasted as usual. The result of this arrangement is that the plates are of opposite polarities on their opposite surfaces, consequently in a battery consisting of several pairs they are from the very nature of things in series, and require no connectors. In all other respects this form is bad, as it is almost impossible to prevent the acid leaking from one partition to another. Knowles constructs his plates by enclosing a block of the paste between two perforated plates of inoxidizable alloy.

§ 281. Another form of chloride cell is the *Laurent-Cely*. In this a mixture of chloride of lead and chloride of zinc is fused together and cast into the shape of small buttons. When cold these are washed in water, which dissolves out the chloride of zinc, leaving the chloride of lead in a porous condition. When dried, these buttons are arranged equally over the surface of a mould, and a plate is formed by pouring in a molten alloy of lead and antimony. The frame thus formed holds the buttons in the desired position. To produce the negative plates from these, they are placed in cells containing acidulated water and zinc plates, which are then connected up together. As the result of the galvanic action thus set up, the buttons of chloride of lead are reduced to the condition of spongy metallic lead. Positive plates can be produced from these latter by cautiously heating them in the air, which converts the buttons into oxide of lead. Very similar to this latter is the chloride cell of the American Electric Storage Battery Co. It differs principally in the shape into which the chloride of lead is cast, a square form being adopted, nine such squares being then placed at equal distances apart in a suitable mould and a lead grid or frame cast around them.

§ 282. The *Tudor Accumulator*. This form, which has met with great favour on the Continent, especially in connection with central station work, is manufactured by the Société Anonyme Belge. It differs from the foregoing, inasmuch as the plates are not gridded. Stout sheet lead is passed between heavy groved rollers, which produce deep groves or serrations along its surface. The serrated plates thus produced are coated with a thin layer of peroxide by electrolysis, after which the interstices are filled in with the usual paste of lead oxides. It is claimed that the preliminary electrolyzation prevents the formation of lead sulphate at the junction of the metal and of the active material. The plates are allowed to dry thoroughly, when they are passed between smooth rollers, which has the effect of closing up the groves, thus holding in the active material.

§ 283. The *Chloride* cell of the Electric Storage Battery Co. of Philadelphia is another good type. To obtain the chloride, commercial lead is melted and converted into a fine powder by blowing a steam jet through it. This is dissolved in nitric acid, and then precipitated from the nitrate solution by the addition of hydrochloric acid. The lead chloride thus produced is thoroughly washed to free it from acid, dried, and melted. It is then cast in suitable moulds into the shape of flat square blocks. A given number of these blocks are arranged in a mould, of the size and shape of the plates required, and a leaden frame cast round them, the molten lead flowing in the spaces between each block and binding the whole into one solid plate. The plates thus prepared are placed in a tank containing a solution of zinc chloride, each plate being alternated with a sheet of zinc. Galvanic action is set up; the zinc plates abstract the chlorine from the lead chloride blocks, leaving spongy metallic lead. Any remaining traces of chlorine are removed by putting the plates (after washing) in a vessel containing dilute nitric acid, and then passing a current through. To form the positives from these, they are packed between perforated ebonite boards and formed into peroxide by electrolytic action in dilute sulphuric acid.

§ 284. A novelty in the way of accumulators is the one just suggested (1901) by Edison. It consists of plates of thin sheet steel about $\frac{24}{1000}$ of an inch thick. These are perforated with square apertures or "pockets." These pockets are filled, in the positive plates, with a super-oxide of nickel (*sic*), and in the negatives with finely-divided metallic iron (obtained by special chemical process) mixed with an equal volume of graphite scales. This latter does not enter into chemical reaction, but serves only to increase the conductivity of the mass. The electrolyte is an aqueous solution of caustic potash containing 20 per cent. of potash. The E.M.F. of a freshly charged cell is 1·5 volt. The normal discharge current rate is 8·64 ampères per square foot of active positive or negative surface. The charging and discharging rates are alike; but no injury accrues to plates by more rapid charging. The storage capacity of the cell is about 14 watts hours per pound: in other words, a cell gives enough energy to lift its own weight through a vertical distance of about 7 miles.

CHAPTER X

ON THE CHOICE OF BATTERIES FOR SPECIFIC PURPOSES

§ 285. THE reader who has attentively perused the preceding chapters, will have noticed that among the various cells described, some are distinguished for great constancy; this quality being usually conjoined to a rather low E.M.F. and a somewhat large internal resistance; others by their power of furnishing a large volume of current, at a low E.M.F. for some considerable time; while others, again, though very energetic in action for a short time, and capable of supplying a very heavy current, at a high E.M.F. have little or no constancy; both the E.M.F. and the current in ampères, falling off very rapidly when the circuit is closed. Some, again, have not a particularly high voltage, and are not fitted to give very heavy currents, but being practically quiescent when the circuit is interrupted, and containing substances which depolarize readily, are eminently adapted to given short intermittent currents spread over long intervals of time. It must be evident that in selecting a battery or cell, care must be taken that it shall possess qualifications suited to the work it may be called upon to perform; and in the same way that no one would choose an oven wherein to smelt iron, or a blast furnace for the purpose of baking bread, so a single fluid chromic acid battery would be adjudged as unsuitable for electric bell work, as Leclanché's would be for keeping an arc lamp going.

§ 286. In all cases in which a minute current, at a steady E.M.F., is required *continuously* for a long period of time, as for instance in "closed circuit bell-ringing," or any other closed relay work, where there is continuously a high resistance in circuit, cells of the Daniell type are simply invaluable. If properly and carefully constructed, they may be depended on to work for several months with hardly any attention. If the cells are placed in a position where they are not liable to be moved, or to be subjected to vibration, large cells of the "Gravity" type are eminently satisfactory; and the author has known cases in which electric clocks and electric devices for astronomical work have been kept steadily going for 18 months or two years by means of a single 1-gallon "Gravity" Daniell cell. Sometimes, however, the current is not required *continuously*, but intermittently for short intervals, extending over long periods of time, say a year or 18 months, as for instance in bell-ringing, automatic signalling of water-level, of rises or falls of temperature, etc., the Leclanché and its congeners, the modern "dry cells" are extremely useful. For working the coils which are used to produce the sparks in the ignition plugs for gas and oil engines, and more especially for those used in motor-cars, etc., dry cells present many advantages. They can be constructed to give a fairly heavy current; and as the action of the coil is intermittent, sufficient time elapses between the instant of breaking and making contact (short though this time be) for the cells to recover themselves somewhat; so that it is possible to fire the engine for a journey of two or three hundred miles with a battery of two or three moderate-sized dry cells. Besides this they are compact, cleanly, and not expensive : they are also lighter than accumulators. These latter, however, are superior to dry cells for this purpose, inasmuch as they can supply current at a higher E.M.F. for a much longer time. For lighting purposes, neither Leclanché's nor dry cells are of any practical use, except only for very short intermittent lighting, of low candle power, such as, for example, lighting a small 1 or 2 candle power lamp, for a second or two at a time, to see the hour by a watch; or in a dark cupboard, to light up momentarily, to enable one to get at any article contained therein. For any lighting extending over more than a few seconds at a time they are absolutely useless. Dry cells are sometimes employed to excite small medical coils· They are *not* peculiarly adapted to this kind of work; but if the primary of the coil be *specially* wound with *fine wire* so as to present a fairly high resistance, they may be used for the purpose, provided the coil is not kept working for too long a time, say 15 minutes as a maximum; and even then they require several hours' repose to recover themselves. The best cell for ordinary coil work is undoubtedly one of the single fluid chromic acid type. It is capable of supplying a large current at a fairly high voltage for two or three hours at a time. For very large coils taking heavy currents, such as those used for X ray, or wireless telegraphy work, the accumulator is the best form of cell to employ. The Leitner "Cupron" cell (§ 92) is also very suitable for this class of work. For small medical coils, especially those put up in small cases for the pocket, the Marie Davy sulphate of mercury cell, or the chloride of silver cells known as the De la Rue, Gaiffe, and Skrivanow, are very convenient. Nearly all plating work, whether gilding, silvering, coppering, or nickeling, is now done by the aid of the dynamo; still there are cases in which small articles may be conveniently and cheaply done by batteries. For coppering from acid solutions, the Smee, the Daniell, and the Walker will be found suitable. Unless there is a high resistance in circuit, one cell will be sufficient; but if the resistance is high, two or more cells *in series* may be used. If the work is large, two or more cells coupled in parallel may be required to give sufficient current. If the copper has to be deposited from an alkaline or cyanide solution, a higher E.M.F. will be needed (from 4 to 5 volts), hence either five Daniells, or two cells of the Bunsen type *in series*, will have to be employed. For gilding, the Daniell and the Smee are both efficient and convenient; for silvering, the same cells are also suitable, or if the resistance is high, the Bunsen and its congeners may be employed. For nickeling, which requires an E.M.F. of about 6 volts, three double fluid chromic acid cells, or Bunsens, or Groves, *in series*, are recommended.

For electro-chemical experiments, in which heavy currents at a fairly high voltage are required, such as the reduction of the alkaline metals from their oxides or chlorides, the Grove, the Bunsen, and the chromic acid cells are very suitable. The two former necessitate the use of a properly constructed stink cupboard, with shaft to carry off the nitreous fumes; the latter can be used in the lecture-room, as no noxious products are given off.

Medical men sometimes require fairly heavy currents for cautery work: lifting or plunge cells of the chromic acid type are useful for this purpose, but accumulators, where facilities obtain for charging, are preferable.

For electric lighting, extending over periods exceeding an hour or more, *accumulators* are really the only satisfactory cells. But it is not everywhere that current can be obtained for the purpose of charging accumulators. In these cases, cells of the Lalande Chaperon type, and more especially Leitner's "Cupron" cells, are excellent. They have little internal resistance, have a large capacity, give off no fumes, and are not expensive in work. But the E.M.F. is low; about 0.7^v; so that for every 2 volts marked on the lamps, three cells in series are required. When only short runs are required, such as, for instance, demonstrations with the optical lantern, or any temporary lighting not exceeding an hour, or at most one hour and a half in duration, single fluid chromic acid cells will be found excellent. They give quite 2 volts per cell, and cells of 1 quart capacity have an internal resistance of less than 0.06 ohm, so that they are capable of supplying a large current. Whatever be the number of volts marked on the lamps, half that number of cells will suffice to light them. For small lantern arc lamps, 25 such cells will give satisfactory results for an hour or more. For driving small electro-magnetic models, automatic clock-winders, small fans, or indeed any small electro-magnetic work, in which the magnet-cores are wound to a high resistance, Daniells Cuprons, and Leclanché or dry cells will give satisfaction; the former for continuous, and the latter for intermittent work. Batteries are not suitable for driving large motors Even accumulators are inconvenient owing to their great weight: while the great cost of energy, as obtained from batteries, renders the employment of these almost prohibitive. Still cases may occur in which no other motive power can be utilized, as, for instance, small launches, small navigable flying machines, etc. In these circumstances, *double fluid* chromic acid cells, put up in light celluloid or ebonite cases, are perhaps the only ones that can be used with any degree of success. The working cost (apart from the initial cost of the cells) comes out at about one shilling and sixpence per horse-power hour.

§ 287. In concluding this work, the author begs to acknowledge his great indebtedness to the many electrical firms who have kindly furnished him with illustrations and information, such as the General Electric Co., the International Electric Co., Messrs. Siemens' Bros., Mr. Leitner, the British Battery Co., and Messrs. Le Carbone; also to that indefatigable scientist, Mr. William Shapland of the University College of Wales, Aberystwith, who devoted much time and labour to the collection of interesting matter, and to the prosecution of experiments and tests with the cells described.

INDEX

Richard Clay & Sons, Limited, London & Bungay.

www.ingramcontent.com/pod-product-compliance
Lightning Source LLC
Chambersburg PA
CBHW080822220526
45467CB00008B/2178